First Multi-Campus Meeting on Basic Science 2017

Journal of Physics: Conference Series Volume 1002

Medellin, Colombia
29 September 2017

Editors:

Fernando Duran Florez
Ely Dannier V. Nino

ISBN: 978-1-5108-6127-5
ISSN: 1742-6588

Printed from e-media with permission by:

Curran Associates, Inc.
57 Morehouse Lane
Red Hook, NY 12571

Some format issues inherent in the e-media version may also appear in this print version.

Printed by Curran Associates, Inc. (2018)

For permission requests, please contact the Institute of Physics at the address below.

Institute of Physics
Dirac House, Temple Back
Bristol BS1 6BE UK

Phone: 44 1 17 929 7481
Fax: 44 1 17 920 0979

techtracking@iop.org

Additional copies of this publication are available from:

Curran Associates, Inc.
57 Morehouse Lane
Red Hook, NY 12571 USA
Phone: 845-758-0400
Fax: 845-758-2633
Email: curran@proceedings.com
Web: www.proceedings.com

Table of contents

Volume 1002

First Multi-Campus Meeting on Basic Science

29 September 2017, Medellin, Colombia

Accepted papers received: 23 March 2018
Published online: 5 April 2018

Preface

Papers

General Topics of Physics and Applications (GTPA)

IOP Conf. Series: Journal of Physics: Conf. Series **1002** (2018) 011001 doi:10.1088/1742-6596/1002/1/011001

Preface

Within the framework of the 50 years of the Basic Science Center of the Universidad Pontificia Bolivariana, Colombia, on September 29, 2017, the First Multi-Campus Meeting on Basic Science was commemorated in the city of Medellin, an event that spread the results of research in the area of natural and exact sciences, and their applications in different areas of knowledge. The research was developed by students, professors, and researchers of the research groups attached to the respective units of Science of the Pontifical Bolivarian University.

The First Multi-Campus Meeting on Basic Science was a space for academic, scientific exchange and experiences related to basic sciences. All papers in these Proceedings refer to one of the following topics covered in the First Multi-Campus Meeting on Basic Science: General Topics of Physics and Applications, General Topics of Mathematics and Applications, General Topics of Chemistry and Applications, General Topics of Biology and Applications, Teaching of Natural Sciences and Mathematics, as well as Articulation Between Sciences and Engineering. The editors hope that those interested in the area of science, can to enjoy this reading, which reflects a wide variety of current issues.

The First Multi-Campus Meeting on Basic Science consisted of 7 Magisterial Conferences presented by professors of the Department of Physics of Southern Illinois University Edwardsville (Edwardsville, United States of America), Aalto University (Helsinki, Finland), Universidad Nacional de Colombia (Medellín, Colombia), Universidad de Antioquia (Medellín, Colombia), Universidad de Medellín (Medellín, Colombia), Politécnico Colombiano Jaime Isaza Cadavid (Medellín, Colombia) and a member of the Optical Society of America. Besides, 35 research work was presented as Oral or Poster by undergraduate and graduate students, professors and researchers from different cities of Colombia.

On behalf of the organizing committee of the First Multi-Campus Meeting on Basic Science, we are extremely thankful to all authors and participants for providing their valuable contributions for these Proceedings as well as the reviewers for their constructive recommendations and criticism aiding to improve the presented articles.

The Editors

Fernando Durán Flórez
Ely Dannier V. Niño

IOP Conf. Series: Journal of Physics: Conf. Series **1002** (2018) 011001 doi:10.1088/1742-6596/1002/1/011001

Organizing Committee
Freddy Rafael Pérez (Chairman)
Claudia Paulina González
Paula Andrea Gutiérrez
Luz Dary Castellanos Prada
María Luz Aída Sabogal Tamayo
Hermes Barrera Ávila
Fernando Durán Flórez
Ricardo Posada Jaramillo

Academic and Scientific Committee
Ph.D. Wilber Silva López, UPB Medellín
Ph.D. Freddy Rafael Pérez, UPB Medellín
Ph.D. Juan Serna Restrepo, UPB Medellín
Ph.D. Héctor Lorduy Gómez, UPB Medellín
Ph.D. Johnson Garzón Reyes, UPB Medellín
Ph.D. Ángel Salazar Martínez, UPB Medellín
Ph.D. Juan Carlos Vásquez Betancur, UPB Medellín
Ph.D. Diego Alejandro Muñoz Durango, UPB Medellín
Ph.D. Claudia Paulina González, UPB Bucaramanga
M.Sc. Luz Dary Castellanos Prada, UPB Medellín
M.Sc. Fernando Durán Flórez, UPB Bucaramanga
M.Sc. Pedro Elías Vera Bautista, UPB Bucaramanga
M.Sc. José Maria Naranjo,UPB Montería
M.Sc. Sixto Bermudez Roa, UPB Montería
M.Sc. Saúl David Bertel Hoyos, UPB Montería
M.Sc. José María Naranjo Martínez, UPB Montería
M.Sc. Johana Paola Rodiño Arguello, UPB Montería
M.Sc. Carlos Alfonso Barrios Villadiego, UPB Montería

IOP Conf. Series: Journal of Physics: Conf. Series **1002** (2018) 011002 doi:10.1088/1742-6596/1002/1/011002

Peer review statement

All papers published in this volume of *Journal of Physics: Conference Series* have been peer reviewed through processes administered by the proceedings Editors. Reviews were conducted by expert referees to the professional and scientific standards expected of a proceedings journal published by IOP Publishing.

Mesoporous silicon oxide films and their uses as templates in obtaining nanostructured conductive polymers

R Salgado[1], G C Arteaga[2] and J M Arias[3]
[1] Corporación Universitaria del Caribe, Sincelejo, Colombia
[2] Universidad del Sinú, Montería, Colombia
[3] Institución Educativa Marceliano Polo, Cereté, Colombia

E-mail: rodrigo.salgado@cecar.edu.co ; rdsalgad@uc.cl

Abstract. Obtaining conductive polymers (CPs) for the manufacture of OLEDs, solar cells, electrochromic devices, sensors, etc., has been possible through the use of electrochemical techniques that allow obtaining films of controlled thickness with positive results in different applications. Current trends point towards the manufacture of nanomaterials, and therefore it is necessary to develop methods that allow obtaining CPs with nanostructured morphology. This is possible by using a porous template to allow the growth of the polymeric materials. However, prior and subsequent treatments are required to separate the material from the template so that it can be evaluated in the applications mentioned above. This is why mesoporous silicon oxide films (template) are essential for the synthesis of nanostructured polymers since both the template and the polymer are obtained on the electrode surface, and therefore it is not necessary to separate the material from the template. Thus, the material can be evaluated directly in the applications mentioned above. The dimensions of the resulting nanostructures will depend on the power, time and technique used for electropolymerization as well as the monomer and the surfactant of the mesoporous film.

1. Introduction
Recently, nanostructured materials from conductive polymers have attracted special attention from disciplines related to science and engineering due to a number of factors such as their large surface area, transport properties, chemical specificity, low cost, ease of processing and large-scale production, which makes them the main candidates to replace materials in the production of micro/nano devices, as well as chemical and biological sensors.

These new morphologies not only reduce the dimensions of the polymeric materials but also contribute to improving their properties for applications such as the manufacture of OLEDs, solar cells, electrochromic devices, sensors, etc., which makes it necessary to implement new synthesis methods to develop nanostructures with even smaller dimensions. Some synthesis methods, including Hard Template and Soft Template, have been developed to obtain nanostructured CPs.

1.1. Hard template
Martin was the first explored this method as a way to produce conductive polymer nanofibers, nanotubes and nanowires [1]. This technique is based on the synthesis of the desired material within the pores or channels of a membrane that acts as a template or reactor so that the desired material grows in each pore of the membrane and then the template is removed resulting in materials with a morphology such as that of nanostructures. Commercial Anodic Aluminum Oxide (AAO) and polycarbonate (PC) membranes

are usually used as templates. Porous AAO membranes are generated electrochemically from metal aluminum, and their pores are arranged in the form of a regular hexagonal network, with a density of 10^{11}pores/cm^2 [2]. PC membranes are manufactured using track-etching technology. These membranes have surfaces with pores up to 10nm in diameter and a density of about 10^9pores/cm^2 [3]. However, although they are commercially available, these membranes have a limited number of pore diameters. This technique has the following advantages:

A. It is a widely used method to make metallic nanomaterials, as well as semiconductors and conductive polymers.

B. The diameter of the nanostructures is controlled by the pore size of the membrane used as template, while the length is adjusted by changing the polymerization time. This method allows making nanotubes and nanowires with different diameters.

C. Nanostructures can be generated either chemically or electrochemically by this method, although the chemical process makes it difficult to separate the nanostructured material from the impurities. The membrane is immersed directly in the solution containing the oxidant and the monomer, and then polymerization within the pores of the membrane is generated using a suitable disturbance.

D. Electrochemical deposition is a cleaner technique. This method involves the use of a metallic film that covers the surface of the membrane to carry out the electrochemical polymerization within the membrane pores.

A large number of nanostructures have been developed using both chemical and electrochemical methods. However, the complex membrane production processes, in addition to the subsequent template removal treatments, which can deform or even destroy nanostructures, sometimes limit their use.

1.2. Soft template

Also called self-assembly, it is a simple, powerful technique for the manufacture of nanostructured conductive polymers that uses surfactant-based micelles to confine the polymerization of the conductive polymer in nanometric dimensions. This method is based on the selective control of non-covalent interactions such as hydrogen bonds and Van der Waals forces, which are the driving forces responsible for self-assembly. There are different types of processes to produce conductive polymer nanostructures using self-assembly including the surfactant as soft-template, which stands out as one of the most widely used methods.

1.2.1. Surfactant as soft-template. Surfactants are molecules that form thermodynamically stable aggregates with intrinsically nanoscale dimensions both regarding dissolution and interfaces. The morphology of the surfactant aggregates can be spherical, cylindrical, or a flat bilayer, depending on the type of surfactant used. [4,5]. This technique has been used to make nanowires, nanospheres, and nanotubes. Compared to techniques such as the hard template, this is a relatively simple, low-cost method. However, it also has disadvantages due to the difficulty in separating the desired product.

On the other hand, electrochemical methods have been used to make CP films of controlled thickness according to the electrochemical technique used (potential cyclic voltammetry, fixed) with positive results in different applications as those mentioned above (OLEDs, solar cells, electrochromic devices, sensors, etc.). Chemical and electrochemical methods have also been used. Although good results have been obtained with current CP materials, new studies aim to improve these materials and produce miniaturized device components to manufacture polymers in the form of nanostructures such as nanofibers, nanotubes, and nanowires. Compared to conventional polymer films these materials exhibit improvements in their conductive properties, as well as a higher value in the surface/volume ratio and a decrease in impedance at the electrode-electrolyte interface. Other methods have been evaluated to avoid complex separation processes when using Hard Template and Soft Template techniques. Hence the importance of mesoporous silicon oxide films, which develop on the surface of the working electrode, thus allowing to evaluate and compare their usefulness in different applications concerning conventional polymers.

In recent years, the electrochemical community has become interested in developing techniques for modifying electrodes using oxide films with nanostructured surfaces [5-7]. This process can be carried out through some techniques [5,8-10] especially those using cationic surfactants capable of accumulating on the electrode surface in the form of semi-micelles due to electrostatic attraction [11-13].

The manufacture of oxide films requires the use of an acid solution composed of cationic surfactant and metal salts capable of forming insoluble hydroxides in basic medium. The application of a cathodic potential to electrolyze water and generate OH^- species on the electrode surface will result in the precipitation of the metal oxide in the vicinity of the electrode, which is molded by the physical space occupied by the surfactant semi-micelle [14]. This process is known as surfactant-assisted polycondensation and allows to generate nanoporous metal oxide films [8,15-17]. The channels are formed perpendicular to the electrode and will grow as the potential step is maintained. The pore diameter will depend exclusively on the radius of the semi-micelle, that is, the type of surfactant used.

Recent efforts in this field have resulted in a significant improvement in electrosynthesis of silica or silicon oxide $(SiO_2)_n$ films by using a cationic surfactant (cetyltrimethylammonium bromide) as a template [18-20]. The polycondensation of silica follows the same principle as that of metal oxides. Since silica is practically inert to chemical attacks and solvents and is also resistant to high temperatures, electrodes modified with these nanoporous films can be used for other purposes. On the other hand, factors such as mass transfer and electrode activity are not affected by the presence of this material [21]. In 2007, Walcarius *et al.* [18] were able to produce nanoporous silica films of 2.9nm pore diameter on various electrode surfaces. This type of processes has also been used in analytical electrochemistry [22,23].

Modified electrodes could also be used to form deposits whose morphology is determined by the shape of the nanoporous film, that is, nanowires. This type of structures exhibits very good permeability to oxidation-reduction or redox reactions of electrochemical, nonpolar and charged molecules that have a known response (ferrocene, potassium ferricyanide, etc.).

2. Experimental

The aqueous silica precursor solution was prepared using 5.0×10^{-2}molL^{-1}KNO_3 and 5.0×10^{-4}molL^{-1}HCl. 1 volume of ethanol was added to the solution to form a 50:50 mixture. A 10.0mL aliquot of the solution was taken and poured into a distilling flask together with 0.4187g cetyltrimethylammonium bromide (CTAB) and 760μL tetraethyl-orthosilicate (TEOS). A magnetic stirrer was used to stir the solution constantly for 2.5 hours so that it can be used in electrochemical experiments.

Using a potentiostatic method the silicon oxide film was placed on different electrode types including platinum (Pt) electrodes by applying -0.9V potential for 5s [18,24-26]; conductive glass (CG) electrodes by applying -1.1V potential for 5s [27]; and stainless steel (SS) electrodes, whose optimum potential is currently being analyzed.

3. Results

Platinum, CG and SS electrodes were deposited in a precursor solution. Cyclic voltammetry (CV) was used to carry out potential scanning ranging from the relaxation potential to more negative potentials thus obtaining voltammetric profiles as shown in Figure 1.

Figure 1 shows a decrease in the current after completion of each cycle, which is consistent with the deposition processes described for low conductive materials. Once the range of potentials in which the template-assisted electrodeposition begins is defined (around -700mV to -1200mV) it is necessary to find out the potential and time required for the formation of the template, which ranges from -700mV to -1200mV, and 5 to 10s, respectively. This procedure is carried out through the application of fixed potential (EP). The application of cathodic potential promotes the formation of hydroxyl ions which are necessary to catalyze the polycondensation of the precursors in the silica film. Once the film was obtained, the electrode is washed to prevent the reaction from continuing and then dried in an oven at 130°C for at least 12 hours.

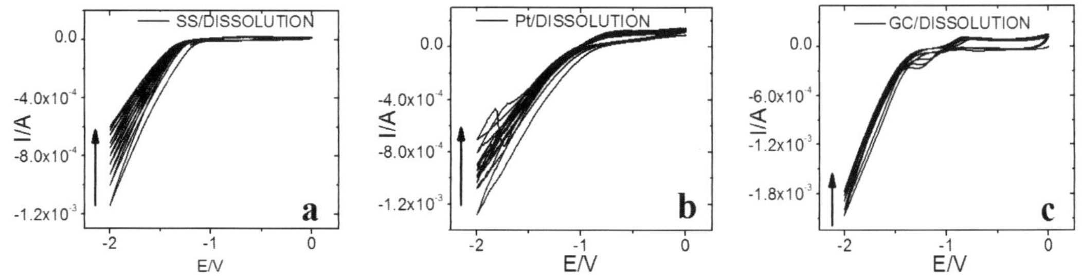

Figure 1. Voltammetry profile of precursor (a) SS|solution, (b) Pt|solution (c) CG|solution. v=100mVs^{-1}. 10 cycles=cycles 2, 5, 7, 9.

Experiments were carried out to demonstrate the permeability of the film due to ion transport. For this purpose, electrodes are immersed in a solution of ferrocene (FeCe) in acetonitrile (ACN). Figure 2 compares the voltammetric responses. It also shows significant reduction of the FeCe current in the modified electrodes.

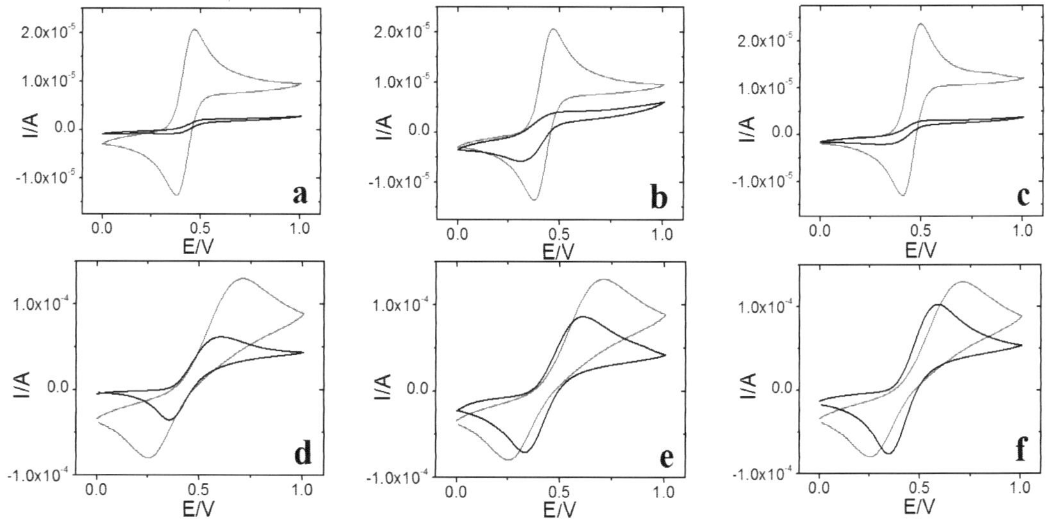

Figure 2. Comparison between the voltammetric responses shows significant reduction of the FeCe current in the modified electrodes Pt|Template and CG|Template. The red line shows the unmodified electrode (a), (b) and (c) =Pt|Template (d), (e) and (f) =CG|Template. The black line shows the modified electrode.

Permeability can be improved by stirring the electrodes in an ethanol-HCl solution which solubilizes CTAB [18] and allows better ion transport through the resulting film. The appropriate potentials for template generation are as follows: -0.9V for Pt, and - 1.1V for CG. This is because mesoporous silicon oxide films exhibit higher permeability in FeCe tests by using these potentials, which suggests that nanostructures can be generated within these films. Tests with template-modified Pt electrodes resulted in the formation of PEDOT [24] and Polythiophene [25] in the form of nanowires, which were later used as specific dopamine (DA) sensors and in electrocatalysis of formic acid [24-26].

Tests with CG electrodes modified with silicon oxide films resulted in polythiophene nanowires [27]. Tests with PEDOT, a conductive polymer with recognized properties, are currently taking place. Experiments are being carried out with SS electrodes to obtain the potential needed for the formation of silicon oxide films in the electrode surface.

4. Conclusions

The modification of electrodes using mesoporous silicon oxide films allows producing conductive polymers, specifically nanowires, with nanostructured morphology. This technique also allows to evaluate the resulting material directly and prevents the material from being deformed by subsequent treatments.

Recommendations

It is advisable to carry out studies using the same type of surfactant with different length chains that promote the formation of templates with larger and smaller pore sizes, which could, therefore, form nanowires of both larger and smaller diameters. This work would allow studying the potential effect of these dimensions on electrocatalysis of elements such as formic acid, and how they can impact the production of DA sensors.

Acknowledgements

The authors would like to thank their respective academic institutions for the support given to their research work.

References

[1] Martin C R, Parthasarathy R, and Menon V 1994 *Electrocim. Acta* **39** 1309-1313
[2] Chakarvarti S K 2009 *Indian J. Phys.* **83(6)** 737-749
[3] Apel P 2001 *Radiat. Meas.* **34** 559–566
[4] Grant L M, Ederth T and Tiberg F 2002 *Langmuir* **16** 2285
[5] Carswell A D W, O'Rear E A and Grady B P 2003 *J. Am. Chem. Soc.* **125** 14793-14800
[6] Zhao D D, Xu M W, Zhou W J, Zhang J and Li H L 2008 *Electrochim. Acta* **53** 2699-2705
[7] Mazaj M, Costacurta S, Zabukovec Logar N, Mali G, Novak Tusar N, Innocenzi P, Malfatti L, Thibault-Starzyk F, Amenitsch H, Kaucic V and Soler-Illia G J A A 2008 *Langmuir* **24** 6220-6225
[8] Tan Y, Srinivasan S and Choi K S 2005 *J. Am. Chem. Soc.* **127** 3596-3604
[9] Newalkar B L, Komarneni S and Katsuki H 2002 *Mater. Lett.* **57** 444-447
[10] Huczko A 2000 *Appl. Phys. A Mater. Sci. Process* **70** 365-376
[11] Thomas A, Goettmann F and Antonietti M 2008 *Chem. Mater.* **20** 738-755
[12] Choi K S and Steinmiller E M P 2008 *Electrochim. Acta* **53** 6953-6960
[13] Sánchez-Rivera A E, Vital-Vaquier V, Romero-Romo M, Ramírez-Silva M T and Palomar- Pardave M 2004 *J. Electrocem. Soc.* **151** C666-C673
[14] Goux A, Etienne M, Aubert E, Lecomte C, Ghanbaja J and Walcarius A 2009 *Chem. Mater.* **21** 731-741
[15] Wang Y D, Ma C L, Wu X H, Sun X D and Li H D 2002 *Talanta* **57** 875-882
[16] Liu P, Lee S H, Tracy C E, Turner J A, Pitts J R and Deb S K 2003 *Solid State Ionics* **165** 223-228
[17] Boeckler C, Oekermann T, Feldhoff A and Wark M 2006 *Langmuir* **22** 9427-9430
[18] Walcarius A, Sibottier E, Etienne M and Ghanbaja J 2007 *Nat. Mater.* **6** 602-608
[19] Etienne M, Quach A, Grosso D, Nicole L, Sánchez C M and Walcarius A 2007 *Chem. Mater.* **19** 844-856
[20] Kwon S Y, Kim J C and Choi B I 2008 *Thin Solid Films* **516** 2843-2847
[21] Etienne M, Sallard S B, Schröder M, Guillemin Y, Mascotto S, Smarsly B M and Walcarius A 2010 *Chem. Mater.* **22** 3426-3432
[22] Walcarius A and Kuhn A 2008 *Trends Anal. Chem.* **27** 593-603
[23] Etienne M, Goux A, Sibottier E and Walcarius A 2009 *J. Nanosci. Nanotechnol.* **9** 2398-2406
[24] del Valle M A, Salgado R and Armijo F 2014 *Int. J. Electrochem. Sci.* **9** 1557-1564
[25] del Valle M A, Ramos A C, Diaz F R and Gacitua M A 2015 *J. Braz. Chem. Soc.* **26** 2313-2320
[26] Salgado R, del Valle M A, Duran B G and Pardo M A, Armijo F 2014 *J. Appl. Electrochem.* **44** 1289-1294
[27] del Valle M A, Gacitúa M, Díaz F R, Armijo F and Río R D 2009 *Electrochem. Commun.* **11** 2117-2120

IOP Conf. Series: Journal of Physics: Conf. Series **1002** (2018) 012002 doi:10.1088/1742-6596/1002/1/012002

Characterization and optimization of an optical and electronic architecture for photon counting

M del M Correa[1] **and F R Pérez**[1]
[1] Universidad Pontificia Bolivariana, Medellín, Colombia

E-mail: maryam.correa@upb.edu.co

Abstract. This work shows a time-domain method for the discrimination and digitization of pulses coming from optical detectors, considering the presence of electronic noise and afterpulsing. The developed signal processing scheme is based on a time-to-digital converter (TDC) and a voltage discriminator. After setting appropriate parameters for taking spectra, acquisition data was corrected by wavelength, intensity response function, and noise suppression. The performance of this scheme is discussed by its characterization as well as the comparison of its spectra to those obtained by an Ocean Optics HR4000 commercial reference.

1. Introduction

There are experiments in which the acquisition of very low luminous intensity is done through optical detectors which have a pulsed current response, such as photomultiplier tubes (PMT), avalanche diodes (APD) or charge-coupled devices (CCD). For PMTs, a pulse often has a nanosecond-order duration and an amplitude of hundreds of millivolts.

The earliest and most used method for signal processing is a dual voltage threshold discriminator and a digital counter [1]. However, this method could not make a distinction between signal and noises that have similar amplitudes and weights, like what happens with afterpulsing and cosmic rays. To make it possible, several schemes have been developed, such as time-to-amplitude converters [2], constant-fraction discriminators [3], simplified schemes of lock-in amplifiers [4], and time-over-threshold discrimination [5]. This last scheme allows a flexible development of a TDC and usage of device built-in LVDS buffers intentionally used as comparators [6]. Moreover, this scheme can use a ramp generator to make a single-slope ADC.

Our aim is to show an alternative to the most used photon counting technique, in order to alleviate the effects of electronic noise and afterpulsing. This work shows a cost-sensitive readout system that offers estimation of pulse parameters, as well as discrimination of amplitude and width. This scheme is targeted at frequency-domain spectroscopy techniques, and it was designed for further demonstration of its usability in time-domain spectroscopy techniques.

2. Discrimination and digitizing scheme set-up

A light source hits a sample. The scattered light is focused by a lens and directed through a slit. An Oriel Cornerstone 130 monochromator filters the passing light by wavelength, which is received by a Hamamatsu R2295 photomultiplier. The output current pulse was preamplified with a Mini-circuits ZX60-P103LN+ module. Then, a LVDS buffer, used intentionally as a comparator, makes a digital representation of the pulse [6-7]. The noninverting output receives the measure signal, whereas the inverting output receives a voltage threshold adjusted by a digital-to-analogue converter (MAX5825,

$1mV$ resolution). A designed and implemented FPGA-based TDC (Xilinx XC6SLX45T, SP-605 board) measures its duration. Its output samples are extracted to characterize the TDC. Also, all samples can be compared with time upper and lower limits in spectra acquisition. If a sample is within these limits, it is considered as a valid count. Finally, the valid counts per second are transferred to the computer in order to plot spectra.

3. Design and implementation of the TDC

The TDC was designed as a set of interpolators receiving multiple clock phases [8-10]. Figure 1 shows a schematic of the developed digital architecture. It is divided in four interpolation phases shifted in quadrature. Each interpolator consists of two fine synchronizers for start and stop signals, a coarse synchronizer, a coarse counter, and two delay lines with their own pre-encoder. The results given by each interpolation phase are pre-processed and coded in a finite state machine. Next, the result goes to a FIFO-memory, which transmits data to RAM memory through AXI-4 DMA core in burst mode. The stream is retrieved by a Microblaze processor and then transmitted to a PC via USB-UART interface. Results follow two paths: one, pulse validation by width and voltage threshold levels in the acquisition of spectra; two, characterization procedures based on the statistical code density test [10]. For this test, 128×103 samples of an uncorrelated $27MHz$ oscillator signal was collected. The results are shown in Figure 2 and Table 1. The test done indicates that DNL (differential non-linearity) varies between $1LSB$ and $2.74LSB$, and INL (integral non-linearity) varies between $-2.75LSB$ and $3.39LSB$. Also, it indicates that the TDC resolution (TLSB) varies between $19.7ps$ and $23.6ps$.

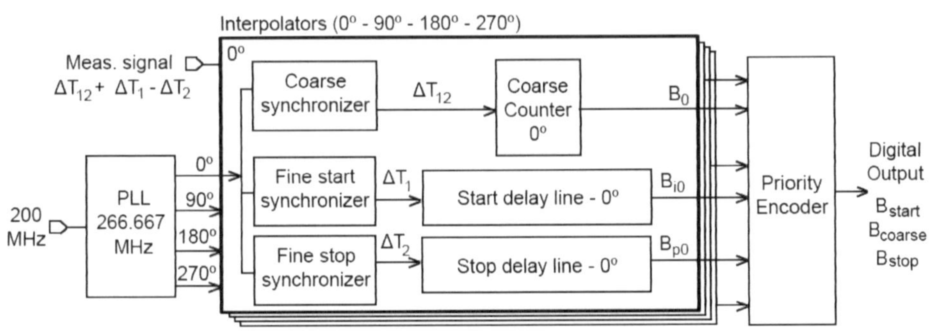

Figure 1. Schematic of the developed digital architecture.

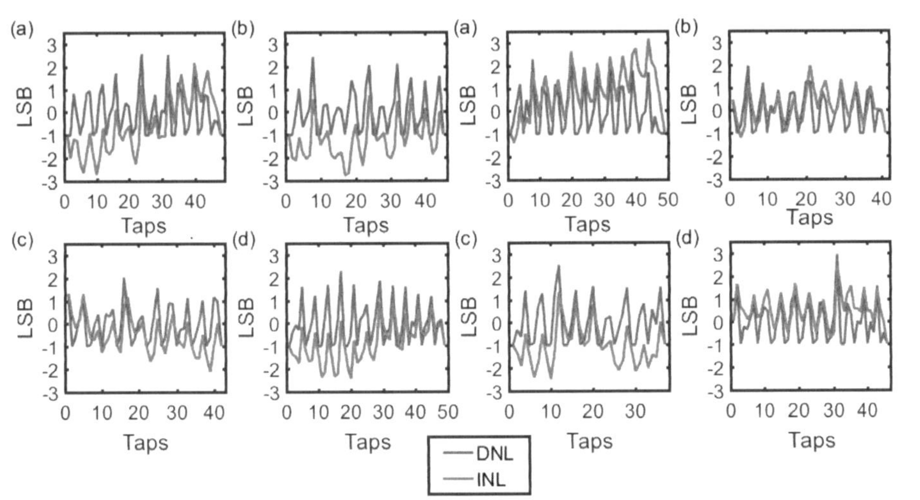

Figure 2. Results of the statistical code density test, showing DNL and INL of start and stop delay lines at (a) 0°, (b), 90°, (c) 180° and (d) 270°, respectively.

IOP Conf. Series: Journal of Physics: Conf. Series **1002** (2018) 012002 doi:10.1088/1742-6596/1002/1/012002

Table 1. Results of the statistical density code test for all delay lines.

Start Lines	Delay line cells	Measurement range (ps)	T_{LSB} (ps)	Stop lines	Delay line cells	Measurement range (ps)	T_{LSB} (ps)
0°	47	923	19.65	0°	48	972	20.69
90°	44	937	21.30	90°	40	895	22.38
180°	41	889	21.68	180°	36	821	22.82
270°	48	999	20.82	270°	45	1060	23.55

4. Optimum time and amplitude parameters

The amplified PMT output signal was captured with a Tektronix TDS2014B oscilloscope in persistence mode, and its graphical plot is shown in Figure 3. The first and main pulse lasts within $2 - 7ns$, which should be consistent with time distributions obtained from the TDC. Also, there is afterpulsing. However, a set of valid time and amplitude limits should reject it. To estimate them, data were taken from time distributions under several voltage thresholds (in Figure 4(a)) and pulse height distributions under a selected time interval (in Figure 4(b)).

In time distributions taken from the TDC, lower voltage thresholds (up to $400mV$) report similar curves, indicating that the pulses from the PMT are being picked up along with afterpulsing and other sources of noise. However, this does not happen for a $500mV$ threshold, in which the shape is scaled, indicating the presence of more counts due to light. For a $600mV$ threshold, the shape is distorted. This is due to the shortening of time intervals generated by the comparator under higher thresholds.

For pulse height distributions, a $2 - 7ns$ time interval was selected. These plots were taken under different light levels, searching for a noise-free level for voltage discrimination. Appropriate voltage levels were found from $525mV$ and beyond. Therefore, a $600mV$ threshold was selected.

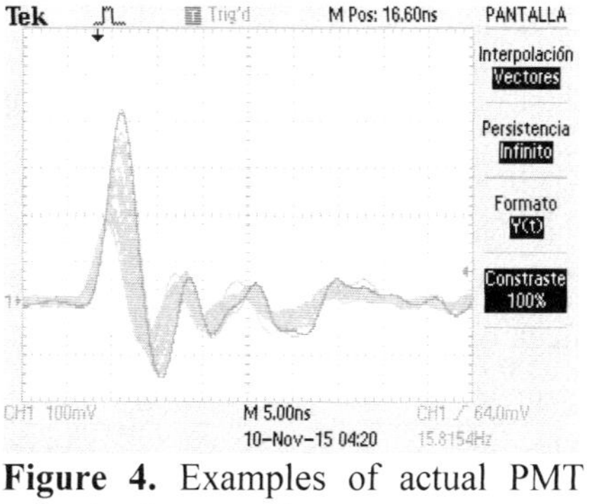

Figure 4. Examples of actual PMT output pulse waveforms.

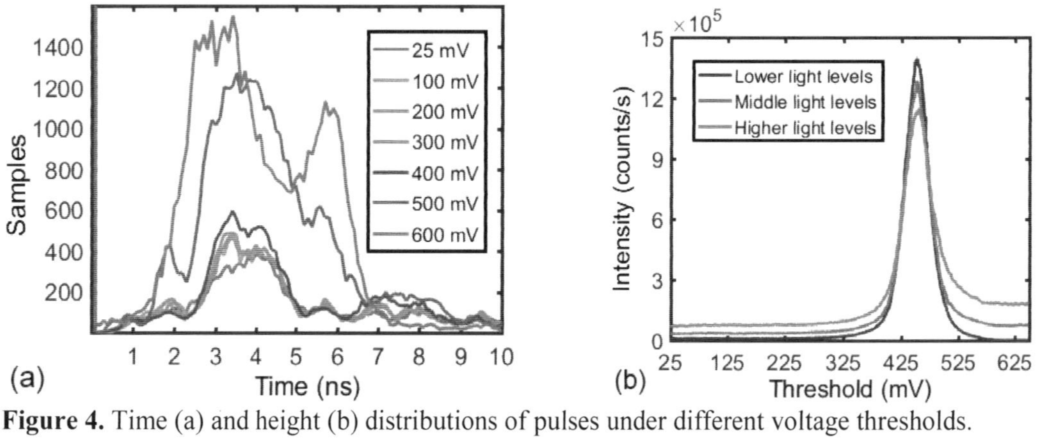

Figure 4. Time (a) and height (b) distributions of pulses under different voltage thresholds.

IOP Conf. Series: Journal of Physics: Conf. Series **1002** (2018) 012002 doi:10.1088/1742-6596/1002/1/012002

5. Acquisition of spectra

After setting the discrimination window, spectra from a mercury lamp and a rhodamine 6G sample in a polyvinyl matrix were taken [11]. In Figures 5 and 6, those spectra are shown, respectively. The raw mercury lamp spectrum was corrected by wavelength and relative intensity, based on the specifications of the optical setup. The fluorescence spectrum was corrected by wavelength only.

In detail, wavelength was corrected through a linear comparison between the mechanical steps from the monochromator and the wavelengths measured for a Nd:YAG and a HeNe laser. Relative intensity was corrected from an instrument response function estimated through a spectrum of a halogen tungsten lamp. Noise suppression and background elimination were done through an algorithm involving moving average filters [12].

The mercury lamp spectrum taken in this work and the respective spectrum taken with the Ocean Optics HR4000 spectrometer cover the main spectral lines. However, there are differences in their position, relative intensity, and background.

For relative intensity, the PMT sensitivity range is from $300nm$ to $650nm$. However, an optimal quantum efficiency ($> 10\%$) is found from $330nm$ to $520nm$. The CCD sensor (Toshiba TCD1304AP) from the spectrometer has a sensitivity range from $200nm$ to $1100nm$. This allows the detection of the spectral line around $312.6nm$ in the CCD sensor.

For position, the most similar spectral lines are around $546nm$, and the $577nm$ and $579nm$ pair, respectively. This does not happen for spectral lines below $500nm$. In this work, it suggests that the wavelength correction algorithm should be refined through a polynomial function.

For background, the raw plot indicates a probability of PMT saturation around $400nm$ and $500nm$, causing changes in heights of spectral lines.

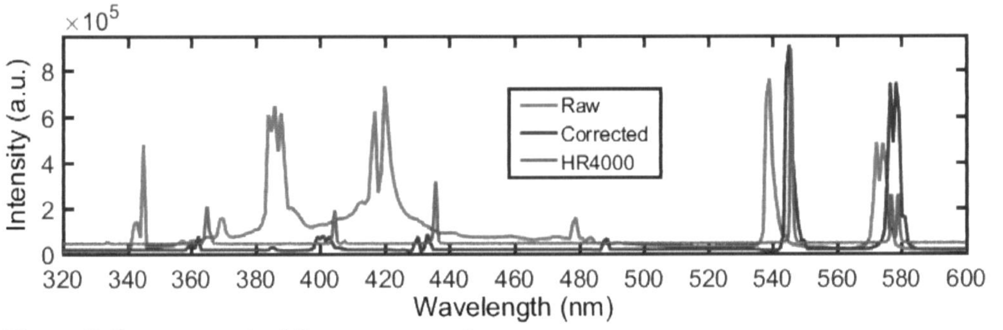

Figure 5. Spectra acquired from a mercury lamp.

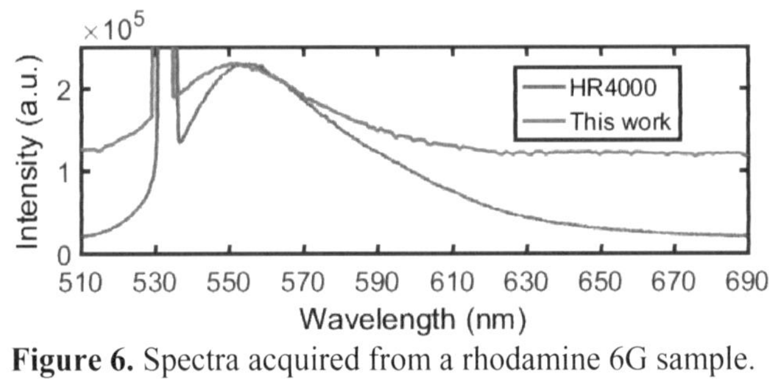

Figure 6. Spectra acquired from a rhodamine 6G sample.

In rhodamine 6G spectra, their shapes are similar. Maxima from this work and the HR4000 reference are $552.8nm$ and $555nm$, respectively. From $620nm$ and beyond, this work cannot retrieve the fluorescence shape, mainly due to sensitivity limitations in the PMT. This means that the choice of the optical detector is essential to ensure the acquisition of broad fluorescence shapes, unlike to what happens in the acquisition of narrow spectral lines.

6. Conclusions

The developed scheme can detect differences between pulses coming from optical detectors and the ones caused by sources of noise, such as after pulsing. This means that it can work under conditions of poor analog signal processing. Also, it can detect differences in levels of light. In conjunction with a TDC and a DAC, the range of pulse height and width of pulses for both optical detectors were found and successfully compared with data obtained from an oscilloscope. These data were used to define discrimination parameters in the acquisition of spectra.

This system will be further refined and used with an optical detector capable of detecting Raman scattering, in order to acquire spectra of low cross-section samples such as samples in powder. In particular, we intend to obtain spectra of samples of carbon nanostructures and iron oxides in the short term.

Acknowledgements

The authors thank Jean-Pierre Galaup from the Laboratory Aimé Cotton at the CNRS and University of Paris Sud-11 for providing some fluorescent samples and his valuable feedback throughout this work. Also, we thank Angel Salazar from Universidad Pontificia Bolivariana for fruitful discussions. We acknowledge the financial support given by Colciencias for the junior researcher grants awarded during 2013 and 2014, and Universidad Pontificia Bolivariana for the scholarship to pursue graduate studies.

References

[1] Becker W 2005 *Advanced time-correlated single photon counting techniques* (Berlin: Springer) p 11
[2] Ripamonti G, Abba A and Geraci A 2010 High frequency, high time resolution time-to-digital converter employing passive resonating circuits *Rev. Sci. Instrum.* **81(5)** 054705
[3] Ghioni M, Cova S, Samori C and Zappa F 1997 True constant fraction trigger circuit for picosecond photon-timing with ultrafast microchannel plate photomultipliers *Rev. Sci. Instrum.* **68(5)** 2228
[4] Restelli A, Abbiati and Gerac A 2005 Digital field programmable gate array-based lock-in amplifier for high-performance photon counting applications *Rev. Sci. Instrum.* **76(9)** 093112
[5] Yonggang W, et al. 2014 A Linear Time-Over-Threshold Digitizing Scheme and Its 64-channel DAQ Prototype Design on FPGA for a Continuous Crystal PET Detector *IEEE Trans. Nucl. Sci.* **61(1)** 99-106
[6] Ugur C, Koening W, Michel J, Palka M and Traxler M 2013 Field programmable gate array based data digitisation with commercial elements *J. Inst.* **8(1)** C91935
[7] Palka M, Moskal P, Bernarski T, et al. 2014 A novel method based solely on field programmable gate array (FPGA) units enabling measurement of time and charge of analog signals in positron emission tomography (PET) *Bio-Algorithms and Med-Systems.* **100(1)** 41-45
[8] Aloisio A, Branchini P, Cicalese R, Giordano R, Izzo V, Loffredo S and Lomoro R 2007 FPGA implementation of a high-resolution time-to-digital converter *2007 IEEE Nucl. Sci. Symp. Conf. Rec.* 504-507
[9] Szplet R, Kalisz J and Jachna Z 2009 A 45 ps time digitizer with a two-phase clock and dual-edge two-stage interpolation in a field programmable gate array device *Meas. Sci. Technol.* **20(2)** 025108
[10] Loffredo S 2010 *Design, construction and tests of a high resolution, high dynamic range time to digital converter* (Roma: Università Degli Studi Roma Tre)
[11] Valdois E 2005 *Elaboration d'echantillons pour applications photophysiques et/ou photochimiques* (Orsay: Université de París Sud-XI)
[12] Baek S J, et al. 2009 A simple background elimination method for Raman spectra *Chemometrics and Intelligent Laboratory Systems* **98(1)** 24

Concretes of low environmental impact obtained by geopolymerization of Metakaolin

D C Sandoval[1], A M Montaño[1], C P González[2] and J Gutiérrez[1]
[1] Universidad Industrial de Santander. Bucaramanga Colombia
[2] Universidad Pontificia Bolivariana, Bucaramanga Colombia

E-mail: claudia.gonzalez@upb.edu.co

Abstract. This work shows results of partial replacement of Portland Type I cement®, by geopolymers obtained through alkaline activation of Metakaolin, in concrete mixtures. Replacement was made with 10%, 20% and 30% of geopolymers at 7, 14, 28 and 90 days of setting. Cement samples was mechanical and electrically tested. Mechanical resistance to compression assay shows that the best percentage of replacement is 10% for every setting time; highest value is 26.75MPa at 90 days. Nyquist diagrams at different times of immersion exhibit same trend: decreasing of electrical resistance as time of assay goes by.

1. Introduction

In concretes, cement is one of the most important elements because of its properties of setting, expansion, and resistance to compression. Nevertheless, in the production cement process, there are huge emission of CO_2, due to calcination of clay and limestone, which in turn, produces clinker. Calcination temperature is so high and to reach it requires uses of petroleum fuels. In last decades, research has been focus on development of new cementitious materials, which production can be environmental friendly. Geopolymerization is a chemical process that can be carried out at room temperature and with reasonable CO_2 emission. Geopolymers are obtained by alkaline agent acting on an aluminosilicate source, these materials have had good cementitious properties [1-4]. This paper shows substitution of cement in concrete mixtures by geopolymers obtained from alkaline activation of Metakaolin [5]. Replacement percentages was 10%, 20% and 30% for setting times of 7, 14, 28 and 90 days. New material formed was tested mechanical and electrically.

2. Experimental

Synthesis of geopolymer started with Metakaolin (Mek) (thermical transformation of commercial Kaolin Faraon), sand, aluminum oxide Al_2O_3 and aluminum silicate Na_2SO_3.mixed until homogeneity. Reaction was performed during $20h$ with continuous stirring at temperature between 60 ℃ and 120 ℃. The mixture is solidified and dried at 60 ℃. Geopolymer (GP Mek) was then crushed according to ASTM C136 [6] to get a particle size of $4.75mm$. New GP Mek, was studied by XRD to find that it is amorphous material with some crystalline phases: muscovite, anatase, microcline and quartz from kaolin starting material.

Concrete specimens were prepared at 10%, 20% and 30% of replacement of Portland by GP Mek and 7, 14, 28 and 90 setting days. Concrete specimens for mechanical strength resistance test had dimensions of $7.5cm$ diameter, $15cm$ high, according to NTC 673 [7]. Specimens for electrical test of $3.0cm$ diameter and $6.0cm$ high, working electrode (steel) and reference electrode (graphite), were submerged in a 2% sodium chloride solution with bubbling oxygen. Design of mixture of concrete followed NTC 3752 specifications [8].

3. Result and discussion

3.1. Mechanical analysis

Results of mechanical assay are shown in Table 1. In this, it can be observed that the specimens with 10% replacement of geopolymer had higher resistances to strength than those for specimens with 20% and 30% for all setting times. In Figure 1, the tendency of the compression resistance values for both the reference specimens and the 10% substitution with geopolymer is shown. Here it can be seen that for 14 and 28 days of setting the resistance values of the specimens with geopolymer substitution are higher than for the reference ones, and even for higher times, the values are very close [9,10].

Table 1. Compressive strength for standard specimen (0% GP Mek), and for specimens of 10%, 20% and 30% of geopolymer (GP Mek) at different setting times.

Age of specimen (setting days)	GP Mek (%)	Compressive strength (MPa)
7	0	10.32
	10	14.76
	20	7.66
	30	5.99
14	0	16.32
	10	17.54
	20	14.10
	30	13.21
28	0	26.19
	10	26.08
	20	23.97
	30	18.76
96	0	27.41
	10	26.75
	20	24.08
	30	19.65

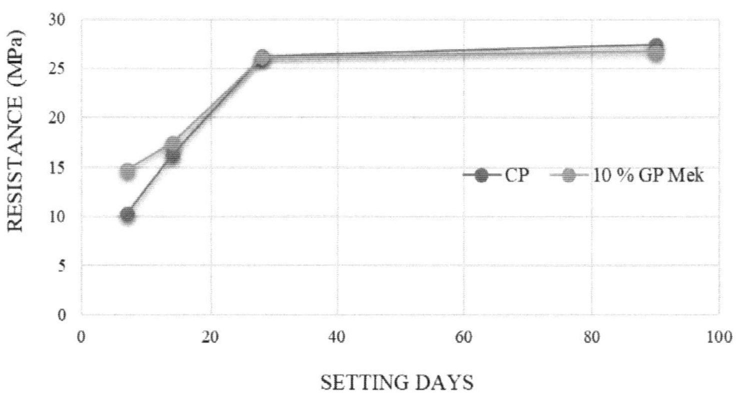

Figure 1. Value of resistance to compression of 10% GP Mek and standard specimen as a function of setting days (The line that joins the points is to see the trend).

3.2. Electrochemical impedance spectroscopy (EIS)

The corrosion behavior of specimens of 90 setting days, with 10% and 20% was studied utilizing electrochemical impedance spectroscopy (IES) technique [11,12]. Specimens with 10% of GP Mek replacement showed highest impedance. Thus it could be more resistant to corrosion.

Figures 2 and 3 shows Nyquist plots for specimens with 10% of GP Mek and standard specimen (0% GP Mek) respectively, at different setting days. As times of study goes by, both plots show increasing impedance. With these diagrams and Bode plots (not shown here), resistance to corrosion (R_Ω), resistance to polarization (R_P) and capacitance (C_p) were calculated (Table 2).

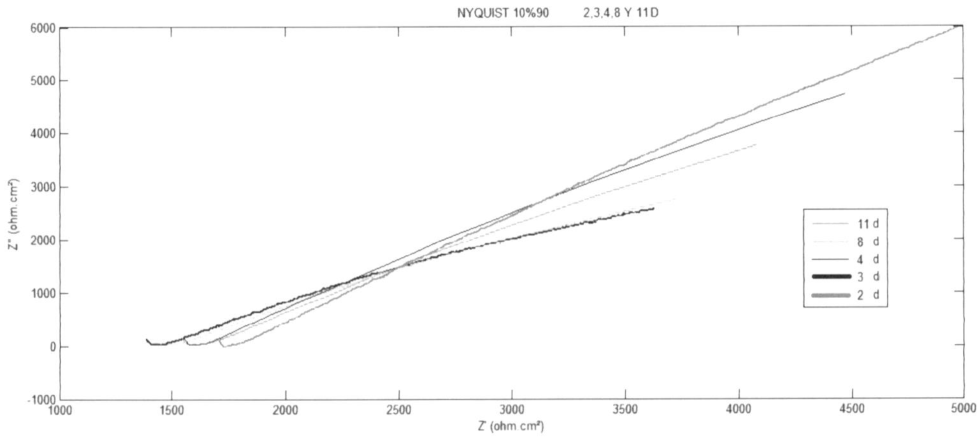

Figure 2. Nyquist plot for 10% GP Mek specimens at different days (show in color).

The values of R_Ω and R_P are high at first day of assay, for both specimens (standard and 10% GP Mek), corroborating thus that there was no corrosion before. For 2 and 11 days of assay, R_Ω and R_P diminish for both kind of specimens, but at second day, 10% GP Mek were lower than the standard specimen resistances. On the eleventh day, the behavior is the opposite. However, on that day, C_p for standard is very low compared to 10% GP Mek, so it is assumed that in the 10% GP Mek, a protective oxide layer has been formed on the steel surface. This oxide decreases the exchange of ions of the solution to the material.

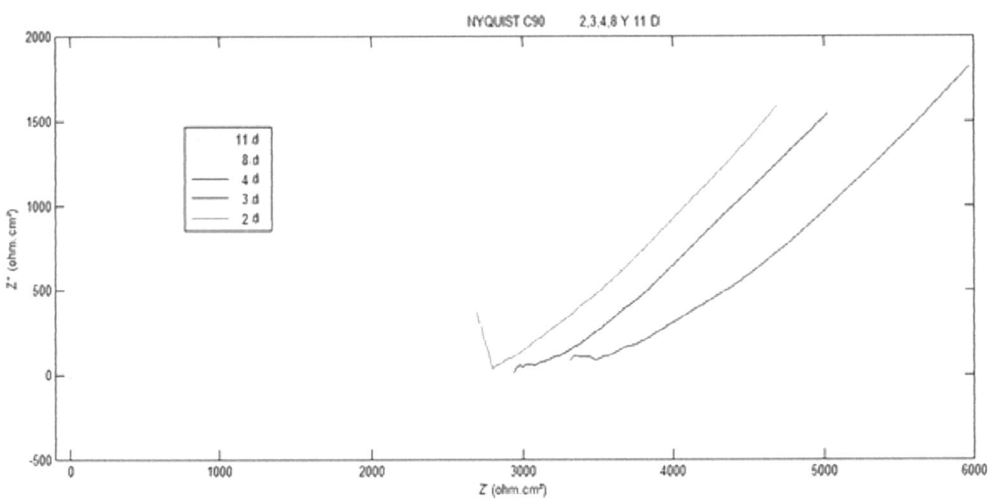

Figure 3. Nyquist plots for standard specimens (0% GP Mek).

Table 2. Values for the simple circuit of some samples evaluates on days 1, 2 and 11.

GP Mek	Day	R_{Ω} (Ω*cm^2)	R_P (Ω*cm^2)	C_P (μF)
	1	1720.00	$3.60x10^{10}$	$2.12x10^{-11}$
10%	2	132.10	3052.1	252.94
	11	128.30	2514.1	307.08
	1	8001.00	$4.17x10^{10}$	$1.85x10^{-11}$
0%	2	369.10	3432.8	224.89
	11	-19.48	38.8	0.21

4. Conclusions

Concrete mixture with replacement of 10% Portland Type I cement by GP Mek, exhibits appropriate mechanical behavior for shorts setting times, thus in the first seven days, it has higher values than those for standard. For large setting times (90 days) mechanical resistance has a slight decrease.

In electrical resistance study, it was found that specimen with 10% of GP Mek has high R_P values, which means a corrosion rate close to zero. With two days of immersion in corrosive medium, R_p is $3052.1\Omega \times cm^2$ and $3432.4\Omega \times cm^2$ for 10% GP Mek and standard, respectively, so rate of corrosion would be lowest for specimen without replacement. Nevertheless, with eleven days, R_P for specimen with replacement is very high yet, which can be due to appearance of protective layer of oxide, according to elevated Cp value found.

References

[1] Sindhunata A, Provis J L, Lukey G, Xu H and Van Deventer J S J 2008 Structural evolution of fly ash based geopolymers in alkaline environments *Ind. Engi. Chem. Res.* **47(9)** 2991-2999
[2] Shayan A, Xu A, Andrews-Phaedonos F 2013 Field applications of geopolymer concrete: a measure towards reducing carbon dioxide emission *Concrete in Australia* **39(3)** 34-44
[3] Martínez C, Mejía J M, Torres J and Mejía R 2015 Evaluation of the toxicity characteristics of two industrial wastes valorized by geopolymerization process *Dyna* **82(190)** 74-81
[4] Komnitsas K and Zaharaki D 2007 Geopolymerisation: A review and prospects for the minerals industry *Minerals Engineering* **20** 1261-1277
[5] Malkawi A B, Nuruddin M F, Fauzi A, Almattarneh H, Mohammed B S 2016 Effects of alkaline solution on properties of the HCFA geopolymer mortars *Procedia Engineering* **148** 710-717
[6] American Society of Testing Materials (ASTM) 2014 *Standard Test Method for Sieve Analysis of Fine and Coarse Aggregates, Standard No. C136/C136M-14* (West Conshohocken: ASTM International)
[7] Instituto Colombiano de Normas Técnicas y Certificación (ICONTEC) 2010 *Ingeniería civil y arquitectura. elaboración y curado de especímenes de concreto para ensayos de laboratorio, Norma Técnica Colombiana NTC 673* (Colombia: Instituto Colombiano de Normas Técnicas y Certificación)
[8] Instituto Colombiano de Normas Técnicas y Certificación (ICONTEC) 2002 *Métodos de ensayo para determinar el contenido de agua en el concreto fresco, Norma Técnica Colombiana NTC 3752* (Colombia: Instituto Colombiano de Normas Técnicas y Certificación)
[9] Wan Q, Rao F, Song S, Cholico-González D F, Ortiz N L 2017 Combination formation in the reinforcement of metakaolin geopolymers with quartz sand *Cem. Concr. Comp.* **80** 115-122
[10] Okoye F N, Durgaprasad J, Singh N B 2015 Mechanical properties of alkali activated fly ash/Kaolin based geopolymer concrete *Constr. Build. Mater.* **98** 685–691
[11] Narayan H, Montaño A M, Hernández M L, Hernández J A, González C P and Ortiz C A 2012 Synthesis, characterization and ac-conductivity measurements of polyaniline based composites with fly-ash and clinker *J. Mater. Environ. Sci.* **3(1)** 137-148
[12] González C P, Montaño A M, González A K, Ríos C A 2014 Synthesis and characterization of inorganic polymers from the alkali activation of an aluminosilicate *IOP Conf. Series: Materials Science and Engineering* **59** 1-5

IOP Conf. Series: Journal of Physics: Conf. Series **1002** (2018) 012004 doi:10.1088/1742-6596/1002/1/012004

Topology optimization applied to the design of cooling channels for plastic injection

D A Muñoz[1], J P Arango[1], C González[1], E Puerto[2] and M Garzón[2]

[1] Universidad Pontificia Bolivariana, Medellín, Colombia
[2] PM-Tec S.A.S. Bogotá, Colombia

E-mail: diego.munoz@upb.edu.co

Abstract. In this paper, topology optimization is applied to design cooling channels in a mold of structural steel. The problem was implemented in COMSOL multiphysics, where two physics were coupled, heat transfer and solid mechanics. The optimization objective is to maximize the conduction heat flux in the mold and minimize the deformations when the plastic is injected. In order to find an optimal geometry for this objective, a density-based method was implemented into the nonlinear program (NLP) for which feasible results were found.

1. Introduction

Nowadays, topology optimization is becoming popular as a design method for thermofluid systems [1]. Traditional designs are based in engineering intuition or in trial and error approaches [2]. Topology optimization brings a non-intuitive design for this kind of systems maximizing or minimizing process variables changing geometric configurations, which affect the physical phenomena. The classic shape of cooling channels for plastic molds are horizontal, parallel or straight ducts because the traditional manufacturing machines. However, this design results in inefficient cooling channels for which heat transfer in the mold is not well removed [3]. Additionally, volumetric contraction, superficial depression or warp can appear causing detrimental mechanical and optical properties [4,5]. On the other hand, long cooling periods translate into low process profitability and limited tool life due to risk of thermal fatigue in the mold [6]. For efficient heat transfer and mechanical resistance in the mold, we use topology optimization with a density-based method for the design of the system [2]. In the density-based method, the algorithm determines the optimal material distribution of a material in a given design domain [1]. In this work, the solid material is distributed in order to maximize heat conduction flux and minimize mechanical deformations with the help of a density field, which takes a value of 1 for solid and 0 for void. This paper is organized as follows. In Section 2 the governing equations are presented. In Section 3, the results are discussed. Finally, in Section 4 we summarize the conclusion of this work.

2. Thermo-mechanic model and governing equations

2.1. Heat transfer modeling

The heat transfer in the solid inside the mold is modeled with equation (1), which consists in the Fourier's law [7]:

$$k\nabla^2 T + Q = 0, \tag{1}$$

Where k is the thermal conductivity of the solid material, T is the temperature, and Q is the heat generation. The objective function related to the heat transfer is shown in equation (2) [7]:

$$A = \int_W k(\nabla T)^2 \, dW, \tag{2}$$

Where W is the design domain, surface or volume. The thermal conductivity is modeled as function of the density variable γ. In equation (3), the material interpolation is shown [7]:

$$k(\gamma) = k_s \gamma^p, \tag{3}$$

Where k_s is the thermal conductivity of the solid phase and p is a penalty factor to improve the convergence of the numerical method. This interpolation allows to decide where the material must appear and where it is not required. The aim is to find the distribution of the control variable to determine the optimal material distribution to improve heat transfer conduction.

2.2. Solid mechanics modeling

The solid mechanics modeling for the deformations in the mold is shown in equation (4), which corresponds to the stresses in the design domain [8] :

$$\nabla \cdot S = 0. \tag{4}$$

The objective function chosen for solid mechanics is shown in equation (5) [8]:

$$B = \int_W Ws \, dW, \tag{5}$$

Where the variable Ws is the deformation energy. The density variable was coupled for solid mechanics through the Young's modulus, E_0, which measures the ability of a material to stretch or compress [8]:

$$E(\gamma) = E_0 \gamma^p. \tag{6}$$

In equation (6), γ is also the control variable and p is other penalty factor to improve the convergence of the numerical method. In this case, the distribution of design variable is computed to keep a mold structure which is able to reduce de deformation when the plastic is injected.

2.3. Objective function

In order to couple both dynamics into a single optimization problem, a multiobjective function for topology optimization is formulated by the sum of equation (2) and (5):

$$O.F = w_1 A + w_2 B, \tag{7}$$

Where w_1 and w_2 are weight factors, taking values between 0 and 1. Note that if one of the factor takes the value of 1, the other part of the objective function is not optimized, for instance, if $w_1 = 1$, only the effect of heat transfer is considered in the topology optimization problem.

3. Results and discussion

In order to validate the problem formulation to design cooling channels in injection plastic molds, two configurations, one 2D and other one 3D, are considered and shown in Figure 1. The 2D system considers a circular chamber where the hot plastic is injected, and the mold will be designed in the square area, while in the 3D system a more complex chamber is considered. The mold material is structural steel, with a thermal conductivity of $44.5W/(mK)$, Young's modulus of $205e^9 Pa$, and the superficial load in the chamber is $100bar$.

In the 2D simulation, a parametric analysis was considered modifying the weigh factors in the objective function, as shown in Figure 1. Note that, for high values of w_2, the fins are designed to decrease deformations without considering the heat transfer [9], as shown in the first row of Figure 2. In contrast, for high values of w_1, the fins are thinner, which increases the heat transfer area favoring the heat removal, as shown in the third row of Figure 2.

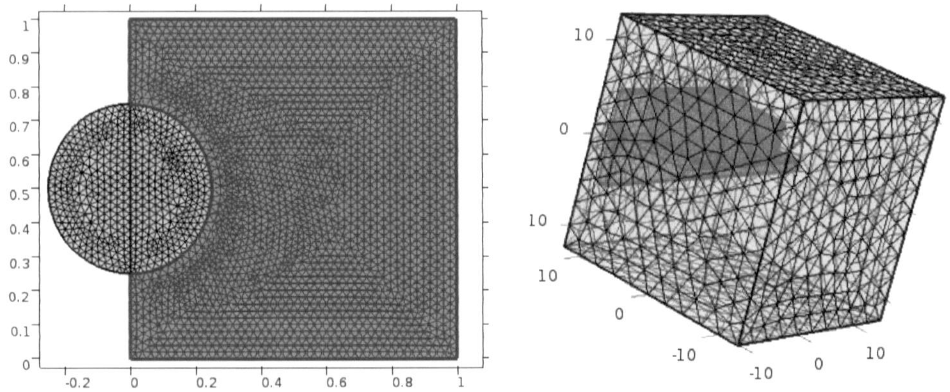

(a) (b)

Figure 1. Geometry of chambers and molds for simulation results. (a) 2D geometry with 4908 finite elements. (b) 3D geometry with 64083.

When both weigh factors have the same input, the density profile γ shows "two big arms" and microchannels round this arms, as shown in the second row of Figure 2. The big arms give to the domain the ability of support high pressures when the plastic is injected, while the thin fins gives to the domain the necessary area to have a good heat exchange between the coolant and the designed mold [10]. For each 2D simulation, 4908 finite elements were considered and the computation time was around 200 minutes, using an Intel Xeon processor of $2.3GHz$ $8GB$ RAM. The weight factors are listed in Table 1. The optimization method was SNOPT for the resulting large-scale nonlinear program.

The results for the 3D domain are presented in Figure 3, where the complete objective function was considered. The red solid parts of the Figure 3(b) are the cooling channels, and the temperature distribution is shown in Figure 3(a). The designed fins give more heat exchange area between the solid part and the coolant. The obtained topology is the one that allows the best configuration. The coolant is going to pass across the mold removing the heat [11]. The weight factors are listed in Table 1. For the 3D domain, 64083 finite elements were considered and the computation time was around 3 days, using an Intel Xeon processor of $2.3GHz$ $8GB$ RAM. The optimization method was also SNOPT for the resulting large-scale nonlinear program.

Finally, Table 1 shows the values of the function objective (7) for every case considered in the simulation results.

Table 1. Objective function value for every case.

Domain	O.F		
2D domain	w_1	w_2	Value
Coupled physics	0.5	0.5	8137.2
Temperature term (K^2)	1	0	298.15
Structure term (J/m)	0	1	4.73e3
3D domain	w_1	w_2	Value
Coupled physics	0.5	0.5	755.1

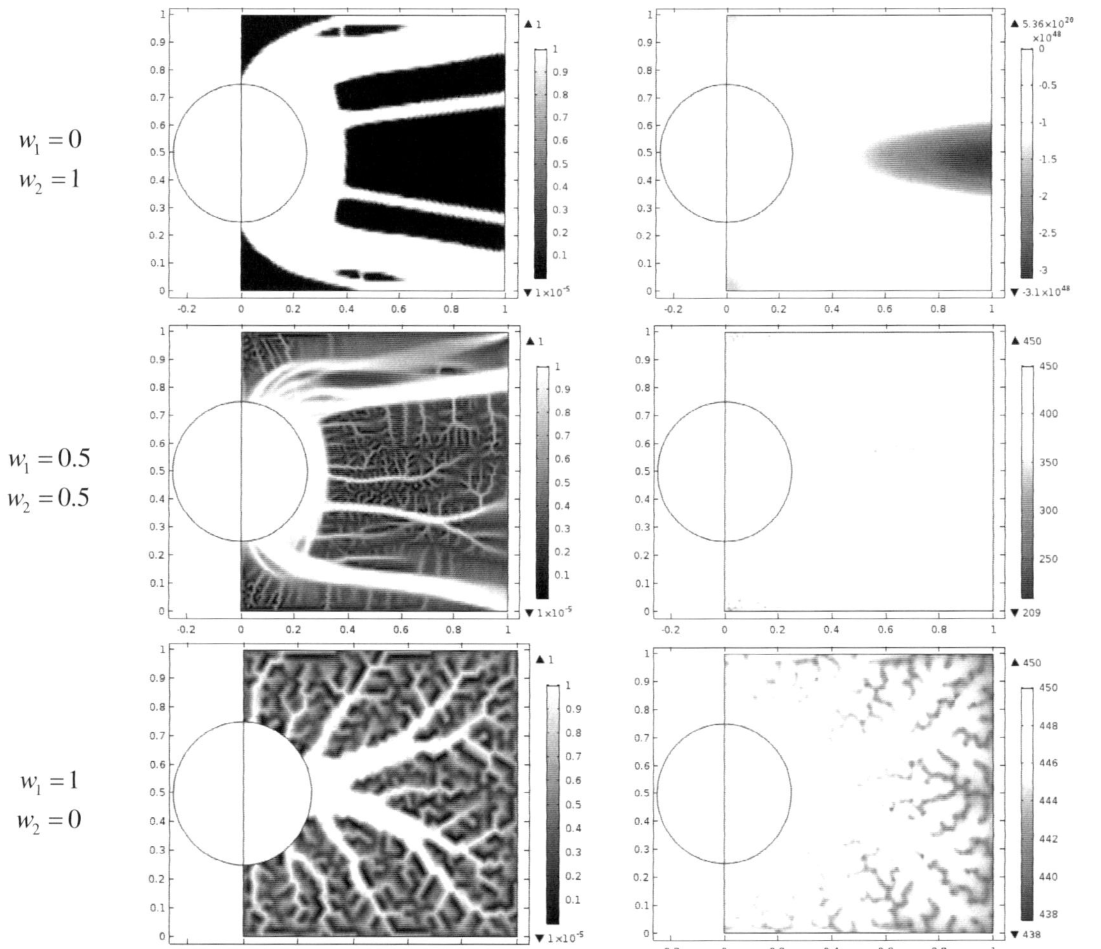

(a) (b)

Figure 2. Parametric analysis for three different sets of weigh factors. (a) simulation results for density profile γ. (b) simulation results for temperature profile.

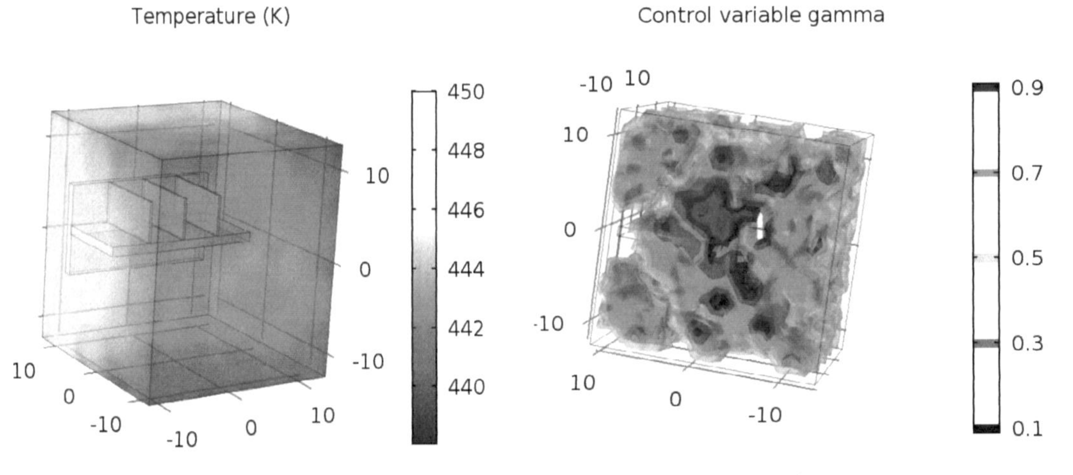

(a) (b)

Figure 3. (a) 3D simulation results for temperature profile. (b) 3D simulation results for density profile γ.

4. Conclusions

The design of molds for plastic injection was formulated using the topology optimization method. Two physics, heat transfer and solid mechanics, coupled with topological optimization allowed to improve the traditional mold manufacture, saving raw material to find the best distribution of material, maximizing the heat transfer and minimizing deformation. The application of topology optimization was solved with the SNOPT method implemented in COMSOL multiphysics. The 3D results show the optimal configuration for the cooling channels for a complex injection chamber. This configuration allows to minimize the temperature in the solid domain guaranteeing structural resistance. The 2D simulation allowed to identify the effect of the weight factors in the objective function.

However, a third physic, the fluid dynamics, should be considered to have a most realistic design. Additionally, a more refined mesh must be considered for the 3D simulation to have a smooth design, which can be manufactured by metal 3D printing [12]. A current work is being developed to handle both problems.

References

[1] Haertel J H K, Engelbrecht K, Lazarov B S and Sigmund O 2015 Topology optimization of therma heat sinks *Proc. COMSOL Conf.* (Grenoble) pp 1-6
[2] Bendsoe M and Sigmund O 2003 *Topology optimization - theory, methods and applications* (Berlin: Springer Verlag) pp 1-158
[3] Xu X, Sachs E and Allen S 2001 The design of conformal cooling channels in injection molding tooling *Polym. Eng. Sci.* **41** 1265-1279
[4] Zhang Y, Huang Z, Zhou H and Li D 2015 A rapid BEM-based method for cooling simulation of injection molding *Eng. Anal. Bound. Elem.* **52** 110–119
[5] Candal M V, Romero J, Müller-karger C, Pelliccioni O 2010 Análisis numérico en el diseño de un molde para prótesis de pie empleando herramientas de diseño (CAD) e ingeniería (CAE) asistidas por computador *Mecánica Computacional* **XXIX** 6441-6454
[6] Sabogal J C 2013 Diseño de la refrigeración en moldes de inyección *Metal Actual* **27** 34–41
[7] Dede E M 2009 Multiphysics topology optimization of heat transfer and fluid flow systems *Proc. COMSOL Conf.* (Boston) pp 1-7
[8] Van Dijk N P, Maute K, Langelaar M and Van Keulen F 2013 Level-set methods for structural topology optimization: A review *Struct. Multidiscip. Optim.* **48** 437–472
[9] Hassani B and Hinton E 2012 *Homogenization and Structural Topology Optimization: Theory, Practice and Software* (London: Springer Science & Business Media) pp 71-101
[10] Raju M and Kumar S 2012 Optimization of heat exchanger designs in metal hydride based hydrogen

storage systems *Int. J. of Hydrogen Energy* **37** 2767-2778

[11] Minkowycz W J, Sparrow E M, Murthy J Y, and Abraham J P 2009 *Handbook of numerical heat transfer* (New York: John Wiley & Sons, Inc) pp 417-442

[12] Cordovilla F, Garzón M, Muñoz D A, Diaz J, García-Beltrán A, and Ocaña J L 2017 A fluid-dynamic numerical model for the selective laser melting of high-thickness metallic layers *Proc. Lasers in Manufacturing Conf.* (Munich) pp 1-10

Morphology of the porous silicon obtained by electrochemical anodization method

S D Bertel H[1], A Dussán C[2] and J M Diaz P[2]
[1] Universidad Pontificia Bolivariana, Montería, Colombia
[2] Universidad Nacional de Colombia, Bogotá, Colombia

E-mail: saul.bertelh@upb.edu.co

Abstract. In this report, the dependence of porous silicon with the synthesis parameters and their correlation with the optical and morphological properties is studied. The P-type silicon-crystalline samples and orientation <1 0 0> were prepared by electrochemical anodization and were characterized using SEM in order to know the evolution of the pore morphology. It was observed that the porosity and thickness of the samples increased with the increase of the concentration in the solution and a high pore density (70%) with a pore size between 40nm and 1.5μm.

1. Introduction

Porous silicon (PS) was initially observed by Uhlir in the 1950s when developing an electrochemical polishing mechanism with hydrofluoric acid (HF) on crystalline silicon wafers to be used in microelectronic circuits [1]. Recently, a wide variety of porous and amorphous materials (including Ti, Al_2, Si, and silica) are investigated for their use on different applications, ranging from optoelectronics to medicine [2-8]. PS, due to its biostability and zero toxicity, is a potentially important material to be considered in many applications including biomedical ones [9,10]. The physical properties of the PS, e.g., electrical resistance, homogeneity in pore formation, thickness, morphology and structural properties, depend on the synthesis parameters during the electrochemical anodization. Thanks to these studies the porous structure is considered as a host matrix for growing hydroxyapatite (HA), thus improving its adhesion and integration with implants and regeneration at the bone level [11-15].

One of the most important properties of HA as a biomaterial for bone regeneration is the dependence with the high pore density of the PS and the possibility of interconnection between them. In this work, PS samples were prepared using as substrate crystalline silicon wafers type-p with <1 0 0> orientation and resistivities varying between $0\Omega cm$ and $100\Omega cm$. The samples were prepared by varying the synthesis parameters such as proportion in the mixing solution, current density, anodizing time, etc.

2. Experimental procedure

Figure 1 shows a schematic diagram of the system used to manufacture PS by the electrochemical anodization method.

HF and Ethanol were obtained from the pharmaceutical company MERCK KGaA, with a purity of 40% and 99.9%, respectively. The substrates used were P-type crystalline silicon wafers (doped with Boron), which had a diameter of approximately $100mm$, crystallographic orientation <1 0 0> and a resistivity between $(0-100)\Omega.cm$. The concentrations in the [HF: C2H5OH] solutions were (1:2),

(1:3) and the currents were varied between $5mA$ and $150mA$, respectively. The reaction times were from $5min$ to $30min$.

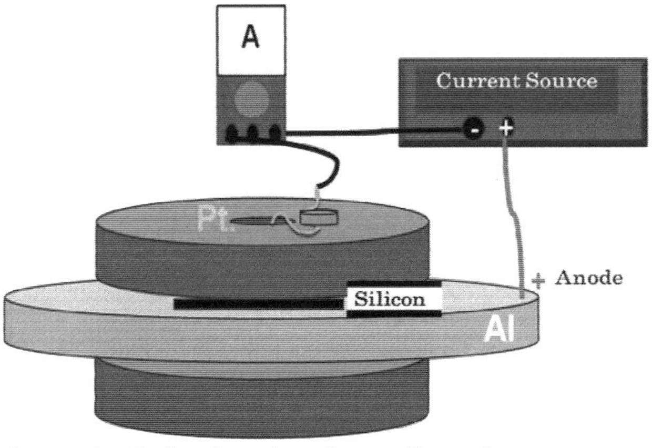

Figure 1. Teflon-based o-ring cell used to carry out electrochemical anodization in the synthesis of PS samples.

The SEM measurements and EDX were performed using a JEOL model JSM 6490-LV microscope. On the other hand, the measures of reflectance were realized with a VARIAN UV-Vis-NIR spectrometer model Cary 5000.

3. Results and discussion

SEM images are presented In the Figure 2 along with the reflectance measurements of PS samples with concentrations (1:2) varying the current density.

It can be seen in the Figure 2 that an increase in the attacking time during the PS synthesis process, influences significantly the formation of grains and non-homogeneous regions on the material surface triggering the loss of the developing silicon porous layer, for $5min$, see Figure 2(a), a current of 5mA incentives the formation of pores with average diameters of $15nm$. The pore density at the surface of the PS was established around 70%, taking into account the current density described by the equation (1).

$$J_{PS} = J\frac{A_{cell}*N_{cell}}{A_{pore}} \qquad (1)$$

where J_{PS} is the current in the pore, A_{pore} is the transversal area of the pore, A_{cell} is a cross-section of a cell of the porous grid and N_{cell} is the number of attacked cells on the silicon surface.

The reflectance measurements obtained for both samples (see Figure 2), are around 25%, with the contribution of reflections corresponding to the near IR being even higher; however, it is possible to observe that the sample made with a time of $30min$ (see Figure 2(b)) presents a greater number of reflections in the ultraviolet region associated to a high absorption and roughness on the surface of the material. Accord to what is observed in Figure 2.

The percentage of reflectance is strictly linked to the pore morphology, for the applications required HA growth of the order of a few hundred nanometers and up to microns has been reported [2,16-22], ergo in case of need of growth of HA the PS porous layers are required to correspond to this order, to promote mechanisms of nucleation of HA within the pore [23].

To study the effect of current density on pore formation, samples of PS were made by maintaining a current of $100mA$ and varying the reaction time.

IOP Conf. Series: Journal of Physics: Conf. Series **1002** (2018) 012005 doi:10.1088/1742-6596/1002/1/012005

(a)

(b)

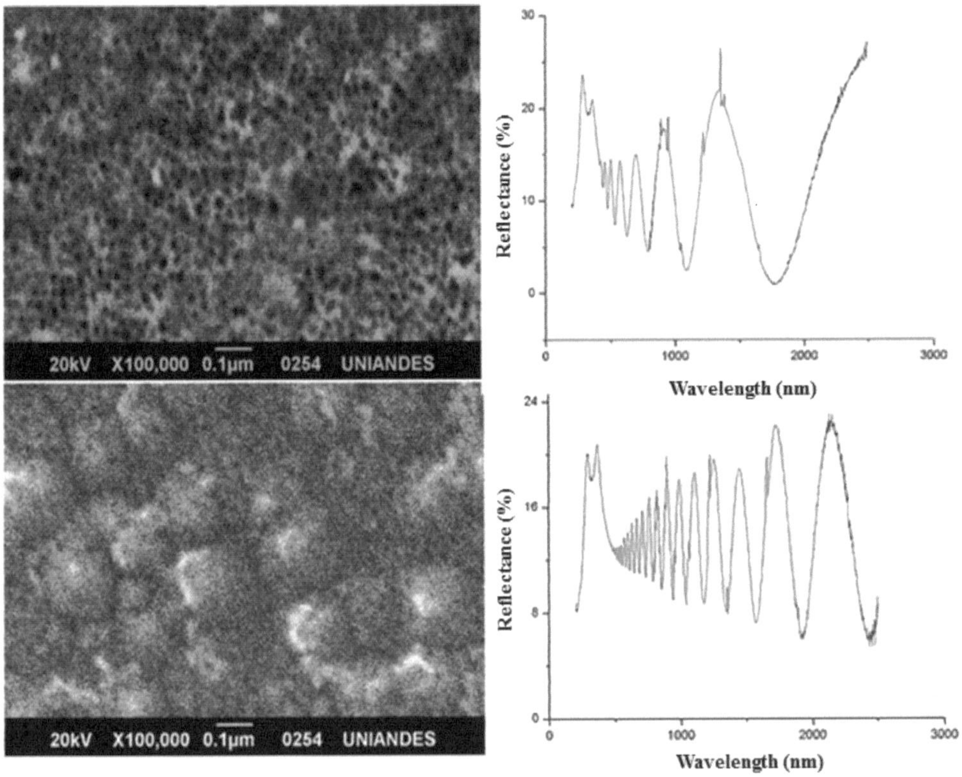

Figure 2. SEM micrograph and reflectance measurements of sample m080711-a obtained with concentration of the mixture (1:2) and current density $5mA$: (a) electrochemical etching time $5min$, (b) $30min$.

A significant increase in the pore diameter of the samples with increasing current at $100mA$ for $10min$ and $20min$ can be observed in Figure 3. By increasing the reaction time from $10min$ to $20min$, keeping both the solution and the current constant, it is observed that the average pore size increases to the order of microns; however it can be seen that the pore density is not completely homogeneous throughout the region. When the applied current was increased to $150mA$, it was possible to observe a pore diameters with various sizes, with a tendency to form pores of diameters up to the order of 4 microns, see Figure 4.

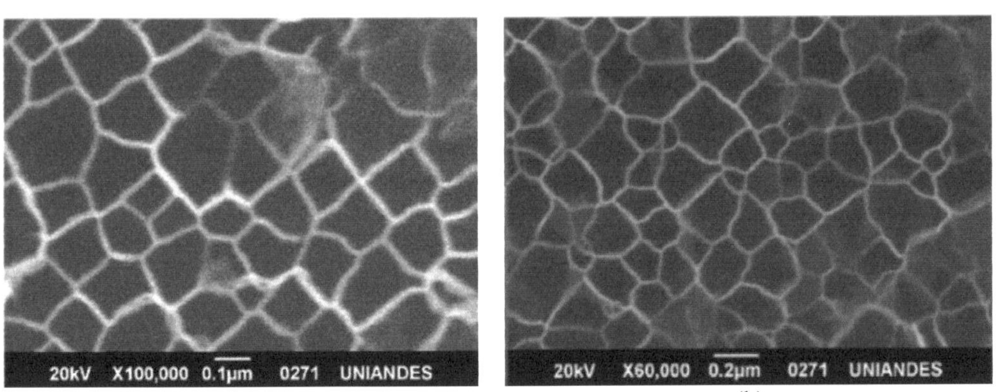

(a) (b)

Figure 3. SEM micrograph of samples of PS attacked (a) during $10min$, with $100mA$ anodizing current and; (b) $20min$ with a current of 150mA.

The sizes obtained could be an interesting feature depending on the application required. However, this result contributes to studies of bone tissue growth using the PS as host substrate [12,24].

The Figure 4(a) shows the SEM micrograph corresponding to the PS sample obtained with a current of $150mA$. Similarly, the Figure 4(b) shows the EDX analysis of the sample in question, showing silicon as a 100% constituent material in the PS formation.

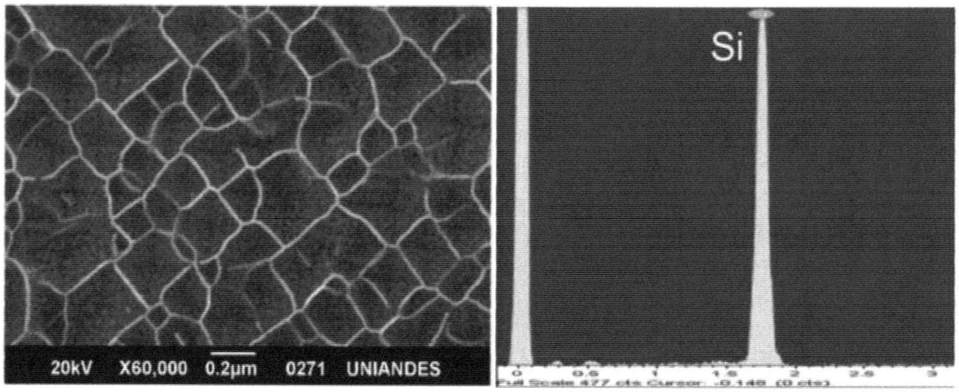

(a) (b)

Figure 4. (a) SEM micrograph of sample m120711-c. (1:2), $150mA$ and $10min$ electrochemical attack. (b) EDX analysis indicating the presence of Si as the predominant material on the surface.

4. Conclusions

PS The porous silicon samples were prepared by the electrochemical anodization by varying the concentration of the mixture, the current and the time of attack during the reaction. It was found that for the concentration (1:2) the reaction time significantly influences the formation of grains and non-homogeneous regions on the surface of the material triggering the loss of the developing silicon porous layer. A high pore density was obtained that varied between 50% and 70%. The pore size varied between $40nm$ and $1.5\mu m$ when the current is increased. The changes in the morphology make PS an interesting material for optoelectronic applications and the growth of hydroxyapatite its use as a bone substitute.

Acknowledgments

This work was supported by Universidad Nacional de Colombia, COLCIENCIAS. We thank the Universidad Pontificia Bolivariana for allowing the research meetings.

References

[1] Uhlir A 1956 Electrolytic shaping of germanium and silicon *Bell System Technical Journal* **35** 333–347
[2] Rhaiti H, Laghzizil A, Saoiabi A, El Asri S, Lahlil K and Gacoin T 2012 Surface properties of porous hydroxyapatite derived from natural phosphate *Mater. Chem. Phys.* **136(2)** 1022–106
[3] Zhou H and Lee J 2011 Nanoscale hydroxyapatite particles for bone tissue engineering *Acta Biomater.* **7(7)** 2769-2781
[4] Nieto A, Hou H, Sailor M, Freeman W and Cheng L 2013 Ocular silicon distribution and clearance following intravitreal injection of porous silicon microparticles *Exp. Eye Res.* **116** 161-168
[5] Wang C, Feng L, Li W, Zheng J, Tian W and Li X 2012 Shape-stabilized phase change materials based on polyethylene glycol/porous carbon composite: the influence of the pore structure of the carbon materials *Sol. Energ. Mat. Sol. C* **105** 21-26
[6] He G, Liu P and Tan Q 2012 Porous titanium materials with entangled wire structure for load-bearing biomedical applications *J. Mech. Behav. Biomed.* **5(1)** 16-31
[7] Vlahinić I, Jennings H, Andrade J and Thomas J 2011 A novel and general form of effective stress in a partially saturated porous material: The influence of microstructure *Mech. Mater.* **43(1)** 25-35

[8] Jianyong W, Huiping T, Yuan G, Baojun Y, Xinting K and Pei W. 2012 Shear behavior on money alloy porous sheet materials *Procedia Engineering* **27** 742-746

[9] Lawrence B, Alagumanikumaran N, Prithivikumaran N, Jeyakumaran N, Ramadas V and Natarajan B 2013 Effect of surface modification on the porous silicon infiltrated with biomolecules *Appl. Surf. Sci.* **264** 767-771

[10] Yokoi K, Godin B, Oborn C, Alexander J, Liu X, Fidler I and Ferrari M 2013 Porous silicon nanocarriers for dual targeting tumor associated endothelial cells and macrophages in stroma of orthotopic human pancreatic cancers *Cancer Lett.* **334(2)** 319-327

[11] Dian J, Macek A, Nižňanský D, Němec I, Vrkoslav V, Chvojka T and Jelınek I 2004 SEM and HRTEM study of porous silicon—relationship between fabrication, morphology and optical properties *Appl. Surf. Sci.* **238(1)** 169-174

[12] Tripathi G and Basu B 2012 A porous hydroxyapatite scaffold for bone tissue engineering: Physico-mechanical and biological evaluations *Ceram. Int.* **38(1)** 341-349

[13] Demetrescu I, Pirvu C and Mitran V 2010 Effect of nano-topographical features of Ti/TiO 2 electrode surface on cell response and electrochemical stability in artificial saliva *Bioelectrochemistry* **79(1)** 122-129

[14] Zhang E and Zou C 2009 Porous titanium and silicon-substituted hydroxyapatite biomodification prepared by a biomimetic process: characterization and in vivo evaluation *Acta Biomater.* **5(5)** 1732-1741

[15] Singh G, Singh H and Sidhu B 2013 Corrosion behavior of plasma sprayed hydroxyapatite and hydroxyapatite-silicon oxide coatings on AISI 304 for biomedical application *Appl. Surf. Sci.* **284** 811-818

[16] Meza D, Figueroa I, Flores-Morales C and Piña-Barba M 2011 Nano hydroxyapatite crystals obtained by colloidal solution *Rev. Mex. Fis.* **57(6)** 471-474

[17] Sequeda L, Díaz J, Gutiérrez S, Perdomo S and Gómez O 2012 Obtención de hidroxiapatita sintética por tres métodos diferentes y su caracterización para ser utilizada como sustituto óseo *Revista Colombiana de Ciencias Químico Farmacéuticas* **41(1)** 50

[18] Feng W, Mu-Sen L, Yu-Peng L and Yong-Xin Q 2005 A simple sol-gel technique for preparing hydroxyapatite nanopowders *Mater. Lett.* **59(8)** 916-919

[19] Liu D, Troczynski T and Tseng W 2001 Water-based sol-gel synthesis of hydroxyapatite: process development *Biomaterials* **22(13)** 1721-1730

[20] Melde B and Stein A 2002 Periodic macroporous hydroxyapatite-containing calcium phosphates *Chem. Mater.* **14(8)** 3326-31

[21] Nayak A 2010 Hydroxyapatite synthesis methodologies: an overview *International Journal of ChemTech Research* **2(2)** 903-907

[22] Aminian A, Solati-Hashjin M, Samadikuchaksaraei A, Bakhshi F, Gorjipour F, Farzadi A and Schmücker M 2011 Synthesis of silicon-substituted hydroxyapatite by a hydrothermal method with two different phosphorous sources *Ceram. Int.* **37(4)** 1219-1229

[23] Gross K, Muller D, Lucas H and Haynes D 2012 Osteoclast resorption of thermal spray hydroxyapatite coatings is influenced by surface topography *Acta biomaterialia* **8(5)** 1948-1956

[24] Pimenta L, Pesántez C and Oliveira L 2008 Silicon matrix calcium phosphate as a bone substitute: early clinical and radiological results in a prospective study with 12-month follow-up *SAS Journal* **2(2)** 62-68

IOP Conf. Series: Journal of Physics: Conf. Series **1002** (2018) 012006 doi:10.1088/1742-6596/1002/1/012006

Design of open rectangular and trapezoidal channels

C P González[1], P E Vera[1], G Carrillo[1] and S García[1]
[1] Universidad Pontificia Bolivariana, Bucaramanga, Colombia

E-mail: claudia.gonzalez@upb.edu.co

Abstract. In this work, the results of designing open channels in rectangular and trapezoidal form are presented. For the development of the same important aspects were taken as determination of flows by means of formula of the rational method, area of the surface for its implementation, optimal form of the flow to meet the needs of that environment. In the design the parameter of the hydraulic radius expressed in terms of the hydraulic area and wet perimeter was determined, considering that the surface on which the fluid flows is the product of the perimeter of the section and the length of the channel and where shear is generated by the condition of no slippage.

1. Introduction

The use of surface drainage systems is of great importance within the scope of civil engineering because it provides stabilization solutions on areas affected by the adverse effects of water flow, which are captured, transported and unloaded at specific points as sources. natural under optimum hydraulic conditions [1-2].

That is why in hydraulic there are devices such as landfills. The design and construction of these is of great importance for the experimental and practical teaching of the engineering student since it brings him closer to a better understanding of its operation for which he needs to determine: flow meters, as well as the limits of application depending on the ground conditions or place of execution.

In this paper, the results of analyzing thin-walled landfills in rectangular and trapezoidal section design, determination of flow rates using hydraulic formulas are reported.

2. Mathematical development

2.1. Calculation of the flow

The flow calculation was determined by means of the formula of the rational method.

$$Q = C \times I \times A \quad (1)$$

Where, Q=Flow (L/s), C=Runoff coefficient, is based on the soil and a value of 0.3 corresponding to slopes with vegetation according to the table D.4.5 RAS 2000 [3-4]. The I=Intensity (L/(s×ha)), is established by curves IDF (Intensity, Duration, and Frequency) [5] for a given return period, assuming a value of $361.11(L/(s \times ha)$ A=Area (ha), which corresponds to the surface exposed to rainfall and where surface runoff will occur. Expressing the equation (1) depending on the drainage area:

$$Q = dv/dt = C \times I \times dA \quad (2)$$

Where, $A = x \times y$. Starting from a change ratio of the surface area of a rectangle with the conditions: $x = 4000m$, $y = 6000m$, possible area where the canal will be built.

$$\frac{dx}{dt} = -\frac{1cm}{s}, \quad \frac{dy}{dt} = \frac{10cm}{s}, \quad \frac{dA}{dt} = y\frac{dx}{dt} + x\frac{dy}{dt} = 340m^2. \tag{3}$$

This differential equation area (3) is integrated for an interval equal to t_2=1s y t_1=0s.

$$\int dA = \int_0^1 340\frac{m^2}{s}dt = 340m^2 \tag{4}$$

Thus, the formula of the flow is:

$$Q = 3.68 * 10^{-3}\ m^3/s \tag{5}$$

2.2. Calculation of the cross-section of the channel with rectangular geometry

The rectangular landfill is one of the simplest to build and for this reason is one of the most used. It is a landfill with a section of flow in the form of rectangle with thin walls, metal, wood or some resistant polymer, with a beveled or cut crest in slope, in order to obtain a thin edge. The accuracy of the reading it offers is determined by its error level ranging between 3 and 5% [6-7]. To determine the dimensions of the section of the channel, we use the continuity formula:

$$Q = A \times V \tag{6}$$

The Manning equation:

$$V = \frac{1m^{(1/3)}}{n} \times (R^{(2/3)}) \times (S^{(1/2)}) \tag{7}$$

Where, R=Hydraulic radio (Hydraulic area/Wet perimeter), S=Channel slope, n=Coefficient of roughness. Assuming the following variables as initial conditions: $b = 15cm$, $n = 0.009$, $S = 0.07\%$. We have:

$$A = b \times yn, \tag{8}$$

The objective equation is established.

$$x^{(5/2)} - 0.01x - 7.62 \times 10^{-4} = 0 \tag{9}$$

To estimate X, the Newton Rapshon iterations method is applied, using the following equation:

$$X_{n+1} = X_n - \frac{f(x)}{f'(x)} \tag{10}$$

Table 1 reports the interactions found by the Newton Rapshon iterations Method. It was evidenced that the variable X of all the evaluated ones closest to generating the equality is presented in the point is $X = 0.0743$

Table 1 Interaction found by the Newton Rapshon.

No	X	f(x) X^(5/2)	g(x) 0.01X+7.62*10^-4	h(x)=f(x)-g(x)
1	0	0	0.0007620	-0.000762
2	0.010	0.00001	0.0008620	-0.000852
3	0.020	5.6569E-05	0.0009620	-0.000905
4	0.030	0.00016	0.0010620	-0.000906
5	0.040	0.00032	0.0011620	-0.000842
6	0.050	0.00056	0.0012620	-0.000703
7	0.060	0.00088	0.0013620	-0.000480
8	0.070	0.00130	0.0014620	-0.000166
9	0.071	0.00134	0.0014720	-0.000129
10	0.072	0.00139	0.0014820	-9.09868E-05
11	0.073	0.00144	0.0014920	-5.21835E-05
12	0.074	0.00149	0.0015020	-1.23669E-05
13	0.0741	0.00149	0.0015030	-8.32930E-06
14	0.0742	0.00150	0.0015040	-4.28145E-06
15	0.0743	0.00150	0.0015050	-2.23382E-07
16	0.076	0.00159	0.0015220	7.03333E-05
17	0.077	0.00165	0.0015320	0.000113
18	0.078	0.00170	0.0015420	0.000157
19	0.079	0.00175	0.0015520	0.000202
20	0.080	0.00181	0.0015620	0.000248

Demonstrated that the value of the fifth iteration satisfies the equality it is determined that the height of the sheet of water (y_n) of the channel for the proposed conditions is 7.40cm and it assumes a free edge of 2.60cm. In the Figure 1, a scheme with the dimensions of the rectangular design is presented.

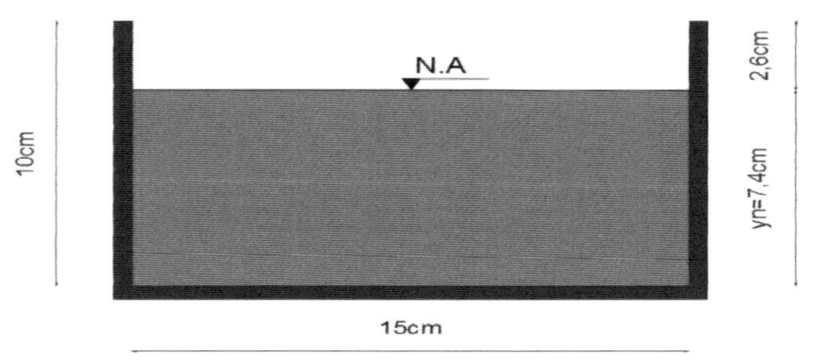

Figure 1. Rectangular design of the channel.

2.2.1. Type of flow. Fluid flows are classified by two types of subcritical and supercritical regimes, for the flow under study was determined under the conditions [8-9]:

$$Sí\ yc > yn\ supercritical\ flow,\ \ Sí\ yc < yn\ subcritical\ flow \tag{12}$$

y_c corresponds to the critical height that will be calculated by the first derivative of the specific energy of the flow as a function of the normal height.

$$Es = Z + Yn + \frac{V^2}{2g} = 0.0755m \tag{13}$$

Where, Z=Load by position, Y=Pressure loading, $V2/2g$: Load by speed: Hydrostatic pressure, g=Gravitational acceleration.

$$V = \frac{Q}{A} \qquad A = b * y \quad for\ rectangular\ section \quad V = 0.33m/s \tag{14}$$

Critical height

$$yc = \sqrt[3]{\frac{(3.68*10^{-3})^2}{(0.15)^2*(9.81)}} = 0.039441m \tag{15}$$

With this critical height value (y_c) the condition is met $Sí\ yc < yn\ subcritical\ flow$ $0.039441m < 0.074m$. Therefore the flow presents a subcritical regime.

2.3. Calculation of the channel with trapezoidal section

A weir with a trapezoidal section compensates for the decrease in flow due to lateral contractions through its triangular parts; this offers the advantage of avoiding correction in the calculations. For the calculation of the trapezoidal section, the dimensions and optimum angle in the elevation of its slopes were determined. In the calculation, the hydraulic area and the constant angle are maintained [10-11].

$$P = b + \frac{2y}{sen\theta}, \qquad A = \left(b + \frac{y}{tan\theta}\right) y \tag{16}$$

Calculation of the elevation angle of the channel slopes.

$$\theta = cos^{-1}\left(\frac{1}{2}\right) \quad \theta = 60^{\underline{o}} \tag{16}$$

Calculation of the trapezoidal section dimensions for the maximum efficiency section with the optimum angle.

$$R = \frac{A}{P} = \frac{y}{2}, \qquad b = \frac{2y(1-cos60^{\underline{o}})}{sen60^{\underline{o}}} = 0.0911m \tag{17}$$

In the Figure 2 shows the dimensions for the section of maximum trapezoidal efficiency.

2.4. Comparison of the maximum rectangular and trapezoidal efficiency section

This parallel is made according to the length of the perimeter of the channel for each of the two sections analyzed, taking into account that an equal channel length is assimilated for the two alternatives [12-13]. Rectangular section perimeter.

$$2(y + c) + b \rightarrow 2(0.074m + 0.026m) + 0.15m = 0.35m \tag{18}$$

Perimeter trapezoidal section.

$$b + \frac{2(y+c)}{sen\theta} \rightarrow 0.09m + \frac{2(0.08m+0.03m)}{sen60^{\underline{o}}} = 0.34m \tag{19}$$

According to this, the implementation of the trapezoidal section is better since it is equivalent to 97.14% of the perimeter of the total rectangular section.

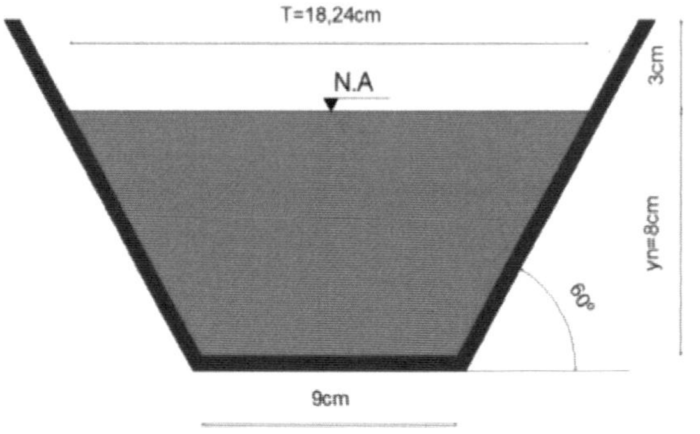

Figure 2 Maximum trapezoidal efficiency section.

3. Conclusions

The method of iterations as a numerical solution is a tool of great importance to find the height of the water sheet of a rectangular section channel for certain operating conditions.

For the cross-section it was evidenced that the variable X of all the evaluated ones closest to generating the equality is presented in the point, verified in the fifth iteration, determining that the height of the water sheet (yn) of the channel for the raised conditions is $7.40cm$ with a free edge of $2.60cm$.

It is determined that the critical height of the rectangular section is $0.039441m$, with which it is obtained that the flow presents a subcritical regime.

According to the calculations made, it was found that the trapezoidal section is the most effective in terms of design and implementation since it is equivalent to 97.14% of the perimeter of the total rectangular section.

References

[1] Chow V 1994 *Hidráulica de canales abiertos* (Santafé de Bogotá: McGraw Hill) pp 513-543
[2] López G, Prado V, Bejamn L, Ruiz C, Carrillo G, Ibáñez C and Arteaga T 2014 *Verificación de procedimientos numéricos de simulación de maniobras en compuertas Vol 5* (México: Instituto México de tecnología del agua) pp 43-56
[3] Saavedra Y and López J 1996 *Comparación de dos esquemas en diferencias finitas para la solución de ecuaciones de flujo no permanente en canales abiertos* (Guayaquil: XII Congreso Latinoamericano de Hidráulica) pp 97-108
[4] De León B 1990 *Modelación matemática de la maniobra de estructuras en una red de canales de riego Vol 4* (México: Instituto México de tecnología del agua) pp 101-109
[5] Giles V 2005 *Mecánica de los fluidos e hidráulica Vol II* (México: McGraw-Hill) pp 45-100
[6] Víctor L, Benjamín W and Keith W 2000 *Mecánica de fluidos* (Santafé de Bogotá: McGraw-Hill) pp 35-80
[7] Burbano J, Juan C and Mustafá I 2002 *Curso experimental de fluidos máquinas hidráulicas* (Pereira: Universidad Tecnológica de Pereira) p 24
[8] Guillermo A 1976 *Manual de hidráulica* (México: Harla) p 578
[9] Ronald V 1969 *Mecánica de los fluidos e hidráulica.* (Mexico: McGraw Hill) p 273
[10] Frederick S 1996 *Manual del ingeniero civil Vol IV* (México: McGraw Hill) p 170
[11] Victor L and Benjamin W 1993 *Mecánica de los fluidos (*México: McGraw Hill) p 595
[12] Máximo V 1995 *Hidráulica de canales* (Costa Rica: Tecnológica de Costa Rica) p 487
[13] Manuel P 1983 *Turbomáquinas hidráulicas* (México: Limusa) pp 37-39

Variational method for calculating the binding energy of the base state of an impurity D⁻ centered on a quantum dot of GaAs-Ga1-xAlxAs

F Durán-Flórez[1]**, L C Caicedo**[1] **and J E Gonzalez**[1]
[1] Universidad Pontificia Bolivariana, Bucaramanga Colombia

E-mail: fernando.duran@upb.edu.co

Abstract. In quantum mechanics it is very difficult to obtain exact solutions, therefore, it is necessary to resort to tools and methods that facilitate the calculations of the solutions of these systems, one of these methods is the variational method that consists in proposing a wave function that depend on several parameters that are adjusted to get close to the exact solution. Authors in the past have performed calculations applying this method using exponential and Gaussian orbital functions with linear and quadratic correlation factors. In this paper, a Gaussian function with a linear correlation factor is proposed, for the calculation of the binding energy of an impurity D^- centered on a quantum dot of radius r, the Gaussian function is dependent on the radius of the quantum dot.

1. Introduction

Nowadays quantum mechanics is an area of physics with a very important boom, since its applications are very diverse, such as, for example, in microelectronics, optoelectronics and other areas where nanotechnology is present [1] and one of the highest concerns is the calculation of link energies, since it guarantees the stability of the quantum system. To study such systems and perform energy calculations, there are many models, methods and tools that allow to address the development of these complex systems, one of these methods is the variational method, where past research has shown the importance of correlation factors for solution of the Hamiltonian [2-5], ensuring the stability of the impurities, in this case, D^- with a confining potential that can be a well or quantum point. In this particular work we seek to calculate the energy of the base state for an impurity D^- centered on a quantum dot of GaAs / Ga1-xAlxAs using as a wave function a Gaussian function as a function of the radius of the quantum dot, with a factor of linear correlation, which have shown their importance when performing calculations supported by the Monte Carlo and variational methods [6-8].

2. Methodology

In the subject of modern physics taught at the Universidad Pontificia Bolivariana based in Bucaramanga, (Santander, Colombia), one of the topics developed is that of quantum mechanics applied to the solid state, beginning by studying the matter waves, the wave duality particle and expected values until arriving at the Hamiltonian approach in one, two and three dimensions. In the class theoretical calculations of simple models are developed, such as the free particle, particle in a box, finite square well and quantum oscillator, but taking advantage of the fact that the university has an Advanced

Computation Center (ACC) they want to use these resources for the calculation of slightly more complex systems such as impurities D^- centered on quantum dots GaAs / Ga1-xAlxAs, for this students are formed under the direction of the teacher who teaches the subject in the programming language Fortram which is the chosen one for the realization of the calculations. The calculation that we want to perform is that of the link energy in this quantum system and for this, it is taken Hamiltonian to a hydrogen model, solving it through the variational method, and for this a Gaussian wave function is proposed that is radio dependent of the quantum dot but where a linear correlation factor is given importance, since in previous studies some authors have shown the importance of these factors to guarantee or ensure the stability of the quantum system [9].

In the work we propose the calculation of the bond energy in the base state of the impurity D^- centered on a spherical quantum dot of radius R of GaAs/Ga1-xAlx, where the wave function is a Gaussian in which takes into account the distances from the center of the quantum dot to the electrons present in the system.

The Hamiltonian for this system is of the form:

$$H = H(r_1) + H(r_2) + 2/r_{12} \quad (1)$$

Were r_{12} is the distance between the electrons and r_1 and r_2 are the distances of each electron to the nucleus. As can be seen, we have the sums of energies present in the quantum system, such as the kinetic and potential energies of the particles and the interaction between them, the Hamiltonian for each of the particles is:

$$H(r_1) = -\nabla_1^2 + V(r_1) - 2/r_1 \quad (2)$$

$$H(r_2) = -\nabla_2^2 + V(r_2) - 2/r_2 \quad (3)$$

Where $V(r)$ is the potential that limits the motion of the particle in two dimensions, and is defined as follows:

$$V = \begin{cases} 0 & r < R \\ V_0 & r > R \end{cases} \quad (4)$$

The proposed wave function as a function of variational parameters is of the form:

$$\psi(r_{12}, r_1, r_2, \theta) = \sum_{i=1}^{2n} A_i(1 + \delta_i r_{12})e^{(-\alpha_i r_1^2 - \beta_i r_2^2)} \quad (5)$$

Where $A_i, \delta_i, \alpha_i, \beta_i$ are the variational parameters that satisfy the cyclic conditions of symmetry:

$$A_{n+1} = A_1 = 1; \quad A_{n+i} = A_i \quad \alpha_{n+i} = \alpha_i; \quad \beta_{n+i} = \beta_i \quad \delta_{n+i} = \delta_i \quad i = 1,2,3, \dots, n \quad (6)$$

When we introduce the wave function into the Hamiltonian, five integrals appear that can be brought into the form of Gamma functions. These are integral energy and correspond to the term of overlap, kinetic energies of the electrons, potential in the barrier, Coulomb interaction with the nucleus and electrostatic repulsion between the electrons which are resolved in the ACC.

3. Results

The binding energies were calculated as a function of the radius of the quantum dot and concordance was observed with other works performed with linear and quadratic correlation factors with exponential wave functions.

It is observed in the data that the energy decreases when the radius of the quantum point increases and increases when it decreases [10]. This surprised the students, since in the analysis of the formula for other systems, such as a quantum well, the energy is inversely proportional to the width of the well, and it would be expected that for very small widths the energy would continue to increase, but not is what the experimental evidences and developed calculations show, something similar happens at the quantum point, it would be expected to continue increasing for very small radii, but the calculations show that the energy has a maximum for a critical radius of approximately 36Å and after this value decreases, this means that the energy changes drastically for values of r less than 36Å, this behaviour is due to the overflow of the wave function, this means that the radius of the quantum point of this value should not be decreased, since the system is affected [11-12]. Some data are shown in the following table:

Table 1. Results of the link energy as a function of the radius of the quantum dot.

R(Å)	$E_b(R_Y^*)$
1.9	0.067
31	2.245
42	2.143
60	1.865
65	1.814
70	1.782
99	1.241

4. Conclusions

The variational method, through history, has proved to be a very efficient method when calculating the link energies for a quantum system and in this particular case it was not the exception, and it was possible to calculate through this, the energy of link of the base state of an impurity $\boldsymbol{D}^-$ centered on a quantum dot of GaAs-Ga1-xAlxAs and the results showed great agreement with other calculations made by other authors in the past.

The maximum energy value was obtained for a critical radius around 36Å, and after this value, the energy drops sharply.

It is observed from the results that for large radii of the quantum dot the binding energy is lower.

The importance of the correlation factors for the stability of the quantum system was evidenced since they take into account the distances and relationships between the particles

References

[1] H A Bethe and E E Salpeter 1957 *Quantum mechanics of one and two electron Atoms* vol 1 (Berlin: Springer-Verlag) chapter 2 pp 118-166

[2] J H Marin, I D Mikhailov, F Garcia 2006 Two electrons in vertically coupled one-dimensional rings *Brazilian Journal Of Physics* **36** 878–881

[3] J H Marín 2005 Transferencia de carga entre donadores en un pozo cuántico *Revista Colombiana de Física* **37(1)** 163-166

[4] J H Marin, I D Mikhailov, F García 2005 Off-axis donors in quasi-two-dimensional quantum dots with cylindrical symmetry *Physica Status Solidi B* **242(8)** 1636-1639

[5] I D Mihailov, F J Betancur and Fernando Durán 1998 Correlación Electrónica en una impureza D- confinada en un punto cuántico *Revista Colombiana de Física* **30(1)** 125-128

[6] T Pang and S G Louie 1990 Negative-donor centers in semiconductors and quantum wells *Phys. Rev. Lett.* **65(13)** 1635-1638

[7] Zhu J L, Zhao J H, Duan W H, Gu B L 1992 D- centers in spherical quantum dots *Phys. Rev. B Condens. Matter.* **46(12)** 7546-7550

[8] Zhu Jia-Lin, Zhao Jie-Hua and Xiong Jia-Jiong 1994 Dimensionality and potential-shape effects on D^0 and D^- ground states in quantum dots *Phys. Rev. B* **50(3)** 1832-1838

[9] F J Betancur, J Sierra-Ortega, R A Escorcia, J D González, I D Mikhailov 2004 Density of Impurity States in Doped Spherical Quantum Dots *Physica E: Low-dimensional Systems and Nanostructures* **23(1-2)** 102-107

[10] J L Zhu, J J Xiong and B L Gu 1990 Confined electron and hydrogenic donor states in a spherical quantum

dot of GaAs- $Ga_{1-x}Al_xAs$ *Physical Review B* **41(9)** 6001-6007

[11] I D Mikhailov, F J Betancur and Fernando Durán 1996 Estado $1s^2$ para una impureza D^- en un punto cuántico *Revista Colombiana de Física* **28(2)** 149-152

[12] I D Mikhailov, FJ Betancur and J H Marín 1996 Intercambio electrónico entre donores hidrogenoides en un pozo cuántico *Revista Colombiana de Física* **28(2)** 157-160

IOP Conf. Series: Journal of Physics: Conf. Series **1002** (2018) 012008 doi:10.1088/1742-6596/1002/1/012008

Physical application: Cannon case

D A Prada[1], **J D Tarazona**[1] **and J M Gómez**[2]
[1] Universidad Pontificia Bolivariana, Floridablanca, Colombia
[2] Universidad Industrial de Santander, Bucaramanga, Colombia

E-mail: duwamg.prada@upb.edu.co

Abstract. The study of physical phenomena by means of guided experimentation and experimental thinking, allow students to infer and understand the reason for the different variations that evidence. Parabolic motion of a projectile powered by a cannon under the spring mechanism, generates discussion regarding the choice of the proper angle, according to a certain distance, a known average initial velocity, and a given height. Give the blank is a great encouragement, however, being able to explain which conditions of the environment influenced the failed launches, generates a space of dialogue and a durable concrete learning.

1. Introduction

The study of the different phenomena that occur in nature has been one of the important activities that are addressed from the Academy. In the halls or classrooms is a theoretical and practical work of some thematic components developed in different university courses. Physics and mathematics are two areas that are closely related, physics as natural science and mathematics as the language to give explanation to the studied phenomena. The rigor that has been given to the teaching of physics, left out its ancestral commitment to unveil the mysteries of the universe and came to be considered only as a new area, at best, that did not correspond to their same expectations and sometimes, with the untiring eagerness to fulfil with the proposed curriculum the enthusiasm of teaching was lost [1]. The importance of generating a lasting knowledge is evidenced when a new information is related to some aspect of what already exists in the cognitive structure of the individual, and thus a process that leads to significant learning is produced [2]. Understand the different variables that affect a case study, allows to create spaces for analysis, discussion, and interpretation that are conducive to a real and meaningful concrete knowledge for those who deal with it. Although equations have traditionally been used as recipes for problem-solving, they provide deeper insight when used as guides to thinking [3].

One of the phenomena of great interest in the study of University Physics is the parabolic motion, for example using cannons. It is important that at the moment of directing this exercise towards meaningful learning, we take into account the pertinent moment, the presentation of the phenomenon, the relationship of the variables, the prediction of the behavior and the environment [4]. The use of compressed air cannons in an undergraduate laboratory provides a way to illustrate the connection between diverse physics concepts, such as conservation of momentum, the work-kinetic energy theorem, gas expansion, air drag, and elementary Newtonian mechanics [5]. Cannons with different specifications, the bullet from the barrel are used in different scenarios, it is possible to use a swab of cotton, in this type of experiment is discussed friction and resistance of the airspeed swab [6]; others focus on the use of compressed air cannon to investigate the impulse, the conservation of the moment and kinematics, also used electronic force plates to measure the recoil of the barrel when fires [7]. In other experiments, meteorological parameters such as wind vector, air pressure, temperature, density and sound velocity are used. [8-11]. There are other types of studies a bit more advanced as the magnetic cannon that converts magnetic energy into kinetic energy. When a steel with low initial speedball hits a string, which consists of a magnet, followed by additional steel balls, the last ball of the chain is expelled to one much larger speed [12]. All these studies enable you to check as seen in class which is out of it. The movement which describes a projectile launched by a cannon, whose mechanism is based on stretching and compression of a spring, can be described with trigonometric equations. However, there are functions, such as the Green function, that are based on the sensitivity models of the projectile trajectory. The best way to analyze the characteristics of the projectiles trajectories under nonstandard conditions is the build of any of the explicit sensitivity models of projectile trajectory [13-16].

Trigonometry as a branch of mathematics is based on the theoretical analysis and practical problems involving triangles basically. For three courses of geometry and trigonometry at the Bolivarian Pontifical University during the two semesters of the academic year 2016, set out as a final project of class, the comparative study of experimental data and data obtained with the trigonometric equations, with which it is possible to calculate the maximum horizontal range of a cell phone shot by a gun, depending on the angle and the initial velocity. The focus is on shooting a small glass, using a cannon powered by spring with some basic specs and General ball, ball glass that should be given a blank ready for a glass of carton with given measures. The distance to which the vessel is left and placed cannons was approximately 2 meters. In order for all the groups that carried out the experiment to be evaluated in the same way, it was necessary to describe the objectives, the competences to be evaluated, the construction parameters, among other data, through a methodological guide and to socialize it from the beginning of the semester

2. Methodological guide of trigonometry project

For the simulation stage of the activity, it is important to stipulate the criterion under well-defined parameters so that the objectives are achieved. The context

2.1. Location

This project will be developed with students from the Universidad Pontificia Bolivariana found studying geometry and trigonometry class. For the development of the same, looking for transversally in each subject of the basic cycle, the student must submit basic knowledge functions, relations, equations.

2.2. Objective

Analyse and argue about the usefulness of the different laws and theorems of trigonometry to represent a model based on a phenomenon of projectile launch.

2.2.1. Specific objectives of the project. (a) Contextualize the concepts learned in geometry and trigonometry class. (b) Design and build a model of canon (projectile based release) with the basic specs. (c) Analyze the launch and (horizontal and vertical) maximum range of a projectile through a canyon constructed, using equations, functions, and trigonometric laws as a tool.

2.3. Justification

Students, using trigonometric ratios, the law of Sines, law of cosines and the trigonometric functions, will analyse various models of phenomena in the environment which can be represented with equations or functions that they involve trigonometric ratios.

2.4. Skills

In the need to assess the members of the groups from their integral formation, it is necessary to stipulate performance indicators for each of the capacities that can be evidenced.

2.4.1. Cognitive skills. (a) Understand parameters analysis and argumentation of trigonometry in response to representations of the context. (b) Analyze the functions and trigonometric reasons as tools that facilitate the solution of various questions in a model that represents a phenomenon of projectile launch.

2.4.2. Communicative competence. (a) Communicate your project smoothly and consistently. (b) Analyze the functions and trigonometric reasons as tools that facilitate the solution of various questions in a model that represents a phenomenon of projectile launch.

2.4.3. Procedural skills. (a) Investigate seriously and gives results according to the provisions. (b) Analyze what are the most suitable design conditions for the optimization of the model. (c) Understand the behavior of the phenomenon according to variability of the environment.

2.4.4. Attitudinal competencies. (a) Values the views of other members of the team. (b) It promotes teamwork. (c) It is consistent with the principles and Bolivarian values of the Universidad Pontificia Bolivariana.

2.5. Design parameters

- Object or mobile: a Mara (small glass marble).
- The use of any type of glue is permitted.
- Must have a base on which rest the object built ($30cm \times 30cm$). About the height is free but should be less than $50cm$.
- The maximum Cannon launch tube $30cm$, the material is free.
- Canon must be fixed to the base and must present an angle for the release meter.
- The force employed for the launch can be generated through a spring or elastic band.
- Canon material, in general, is free.

2.6. The target parameters

- Glass of carton.
- Height $7.6cm$.
- Height inclined $7.7cm$.
- Diameter base lower 5cm, diameter base top approximately $7.5cm$.

Groups known distance to which this vessel at the precise moment of the experiment. With this data, they perform the calculations required to find the angle of launch and proceed to realize three direct releases. Crystal ball releases are marked on a sheet of paper that has with carbon paper. Three is the number of maximum possible launches.

At the end of the experimental activity is the delivery of a written paper type, in which the groups write the corresponding analysis of the observed. At this stage, groups should consider conditions low which the projectile was able or not hit the target. This activity requires an appropriation of the phenomenon in terms of the variables that this involves. In this space the use of experimental thought is evident to understand variables and how these they influenced the observed phenomenon, also this kind of thinking in physics teaching helps students to use their imaginations, to form hypotheses and draw conclusions, all of which are objectives that should be pursued in the modern teaching of Science [17]. Students who attempt to "interpret" mentally an experiment, express their points of view on phenomena that occur in the system under study, allowing teachers to be aware of the ideas of his pupils [17-19].

3. Mathematical method

In a course of University basic physics, the study of the phenomenon of the parabolic motion is made in conjunction with appropriate laboratory. This practice of laboratory and theoretical classes usually are makes it clear that free fall acceleration is constant in the range of movement and is directed downward and in addition, that the effect of the air resistance is negligible. Under these assumptions is that the trajectory of a projectile always describes a parabola. Vector shape describes the function position of the projectile by:

$$\vec{r}_f = \vec{r}_i + \vec{v}_i t + [{}^1/_2]\vec{g}t^2 \tag{1}$$

Where $\vec{r}_f$ represents the final position, $\vec{r}_i$ initial position, $\vec{v}_i$ the initial speed, t the time and $\vec{g}$ the acceleration of gravity.

The components of the initial velocity of the projectile in the flat are determined by:

$$v_{xi} = v_i \cos(\theta_i) \qquad v_{yi} = v_i \operatorname{sen}(\theta_i) \tag{2}$$

Equations (1) and (2) is that the model that describes the phenomenon of parabolic motion, whit respect to its horizontal and vertical scope depends on v_i, θ_i, as seen in Figure 1, and the value of g.

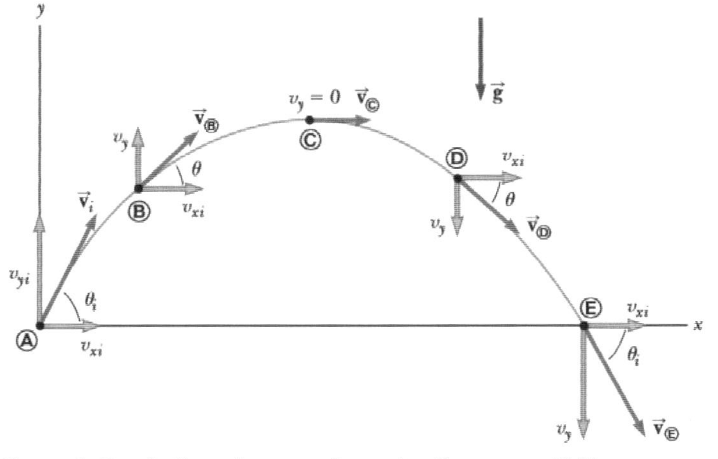

Figure 1. Parabolic trajectory of a projectile, source [20].

The experimental part it is possible to make inferences, scientists like Newton and Einstein, made use of mental experiments to develop, produce, or even to communicate his theories [18,21,22]. In the subject of trigonometry, it was proposed to use the following equations:

$$R = \frac{v_0^2\cos(\theta)}{g}\left[sen(\theta) + \sqrt{sen^2(\theta) + \left[\frac{2gh}{v_0^2}\right]}\right] \tag{3}$$

For which R represents the scope. It is possible to show that if θ satisfies.

$$\cos(2\theta) = \frac{gh}{v_0^2+gh} \tag{4}$$

Then the maximum horizontal range can be determined by:

$$R_{max} = \frac{v_0\sqrt{v_0^2+2gh}}{g} \tag{5}$$

4. Results and discussion

The analysis of parabolic motion using canon, with Spring-based mechanism, by the groups was emphasized in assert that there were other variables such as air friction, the spring elasticity coefficient and wear of the material, which influenced the practical and theoretical calculations in the experiment. Of the 12 groups presented, only 25% (3 groups) they managed to hit the target in three possible attempts, i.e., in one of three attempts, the crystal ball was inside the vessel placed 2 meters from the base of the Canyon, as seen in Figure 2(a). For 36 releases, 3 for each group, the 88% (32 releases) of them hit to $50cm$ of the target approximately, implying that 12% of them hit less than $50cm$ of the target, and the other 8% of the total hit the target, this relation between the releases is observed in Figure 2(b).

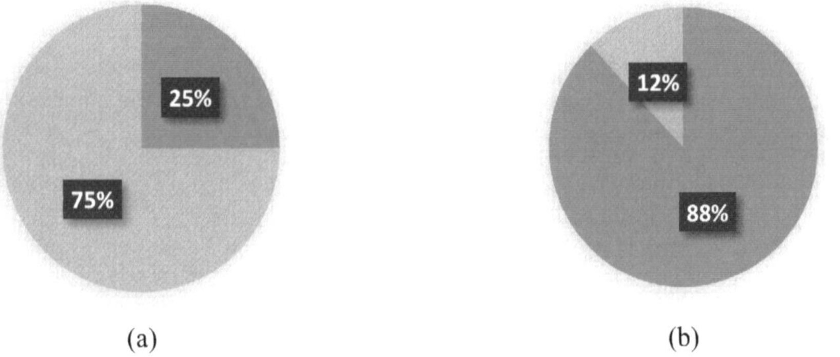

Figure 2. (a) Percentage of hits by groups (b) Percentage of hits by throws.

Average of the horizontal scope with an initial speed varied between $3.15m/s$ and $3.57m/s$, is shown in Table 1, which shows the relationship between angle and distance.

Table 1. The horizontal scope data with respect to the exit angle.

Angle	Distance (cm)
10°	113.000
20°	139.000
30°	169.330
40°	208.160
45°	225.875
50°	183.500

By observing physical phenomena, such as compound movement or projectile movement, the student can identify the variables that affect the phenomenon, since his proposal under the equations presented approaches the evidence in the simulation, which allows him to create a structure of thought under which it adopts analytical positions towards experience and this, in turn, drives it to infer conclusions regarding the same phenomenon [1].

5. Conclusion

With the use of experimental thought, generated a space in which student can mentally interpret an experiment, arguing characteristics which presents a phenomenon in study. The value of the angle for each canon was the same, and in some groups, the difference was considerable respect to the distance reached, because one could conclude that the coefficient of elasticity plays an important role in the calculation of the maximum horizontal distance. Using a strategy to explain two-dimensional movements is a powerful tool for student learning, who find access to knowledge at their fingertips. From the experimental part it can be concluded that the maximum horizontal reach was achieved with the 45° angle and as shown in Table 1 and Figure 2, there are more variables that affect this phenomenon which were not contemplated in the use of the equations proposed, which shows that the failure has a direct relationship with external variables. It was concluded that the equations are models that can represent a phenomenon, but their validity will also depend on the most representative variables that influence the environment.

Reference

[1] Florez M 2011 *Estratégia experimental para la enseñanza del movimiento de proyectiles y el movimiento circular uniforme utilizando el contexto* (Manizales: Universidad Nacional de Colombia)
[2] Novak J 2000 *Aprendiendo a aprender* (Barcelona: Reverté) pp 220-343
[3] Hewitt P 2017 *Sci. Teach.* **84(3)** 14-18
[4] Chavez S and Ortiz G 2008 *Rev. Cam. Abi.* **174** 8-12
[5] Rohrbach Z, Buresh T and Madsen M 2012 *Amer. J. of Phy.* **80** 24-26
[6] Tsukamoto K and Uchino M 2008 *Phy. Teach.* **46** 334-336
[7] Taylor B 2006 *Phy. Teach.* **44** 582-584
[8] Bockarev A, Andrejevskij V and Belokonov V 1985 *Aeromechanics of the aircraft.* Dynamics of flight (Moscow: Mashinostroyenie) p 360
[9] Golubev I, Svetlov V, et al. 2001 *Design of anti-aircraft missiles 2nd ed.* (Moscow: MAI) p 732
[10] Kovalenko V and Shevkunov V 1975 *Meteorological preparation of artillery fire* (Russian: Military Artillery Academy of M. I. Kalinin) p 84
[11] Zipfel P H 2007 *Modeling and simulation of aerospace vehicle dynamics 2nd ed.* (USA: AIAA, Education series) p 607
[12] Chemin A, Besserve P, Caussarieu A, Taberlet N and Plihon N 201 *Amer. J. of Phy.* **85** 495-502
[13] Bliss G 1920 *Trans. of the Amer. Math. Soc.* **21(2)** 93-106
[14] Bliss G 1944 *Mathematics for exterior ballistics* (London: John Wiley and Sons, Inc) p 128
[15] McShane E, Kelley J and Reno F 1953 *Exterior ballistics* (Denver: University of Denver Press) p 834
[16] Molitz H and Strobel R 1963 *Exterior ballistics* (Berlin: Springer-Verlag) p 610
[17] Matthews M 1994 *Thought experiments, Science teaching. The role of history and philosophy of science* (London: Routledge) pp 99–105
[18] Velentzas A and Halkia K 2013 *Sci. Edu.* **22** 2621-2640
[19] Mach E 1976 *On thought experiment, Knowledge and error: Sketches on the psychology of enquiry* (Dordrecht: Reidel Publishing Co) pp 134–147
[20] Serway R and Jewett J 2008 *Física para ciencias e ingeniería vol 7* (México: Cengage Learning) pp 723
[21] Brown J 1991 *The laboratory of the mind. Thought experiments in the natural sciences* (London: Routledge)
[22] Sorensen R 1992 *Thought experiments* (New York: Oxford University Press)

QRS complex detection based on continuous density hidden Markov models using univariate observations

S Sotelo[1], W Arenas[1] and M Altuve[1]

[1] Universidad Pontificia Bolivariana, Floridablanca, Colombia

E-mail: silvia.sotelo@upb.edu.co

Abstract. In the electrocardiogram (ECG), the detection of QRS complexes is a fundamental step in the ECG signal processing chain since it allows the determination of other characteristics waves of the ECG and provides information about heart rate variability. In this work, an automatic QRS complex detector based on continuous density hidden Markov models (HMM) is proposed. HMM were trained using univariate observation sequences taken either from QRS complexes or their derivatives. The detection approach is based on the log-likelihood comparison of the observation sequence with a fixed threshold. A sliding window was used to obtain the observation sequence to be evaluated by the model. The threshold was optimized by receiver operating characteristic curves. Sensitivity (Sen), specificity (Spc) and $F1$ score were used to evaluate the detection performance. The approach was validated using ECG recordings from the MIT-BIH Arrhythmia database. A 6-fold cross-validation shows that the best detection performance was achieved with 2 states HMM trained with QRS complexes sequences ($Sen = 0.668$, $Spc = 0.360$ and $F1 = 0.309$). We concluded that these univariate sequences provide enough information to characterize the QRS complex dynamics from HMM. Future works are directed to the use of multivariate observations to increase the detection performance.

1. Introduction

The electrocardiogram (ECG) signal reflects the electric activity of the heart and is employed as a noninvasive tool for cardiovascular system analysis and the diagnosis of cardiac pathologies. The ECG has a special morphology characterized by the P-wave and the QRS complex, which reflect the atrial and ventricular depolarizations, respectively, and the T-wave, which reflects the ventricular repolarization. In particular, the QRS complex has the largest amplitude of the ECG waves and its detection has a clinical interest due to the possibility to estimate statistical indicators to evaluate the cardiovascular status [1, 2].

Taking advantage of the progress of computing power, different algorithms have been conceived for the automatic detection of the QRS complex. The diversity of the approaches implemented to develop accurate QRS complex detectors goes from digital filters to statistical methods such as wavelet transforms [3], artificial neural networks [4, 5], genetic algorithms [6], mathematical morphology [7], hidden Markov models [2,8,9], among others [10–12]. For instance, algorithms based on the wavelet transform [3] exploit the possibility to analyze the ECG signal at different temporal and frequency resolution scales to distinguish the QRS complex which is particularly characterized by a high frequency component. On the other hand, exploiting the nonlinear properties of artificial neural networks, their parameters have been trained to identify QRS complexes with a better detection performance compared with the linear counterpart [4]. Currently, the performance attained from the implementation of these methodologies are pretty good but questions related with the influence of the signal noise and abnormal morphologies are still open.

A QRS complex detection algorithm based on continuous hidden Markov models is presented in this paper. The models are trained to characterize the dynamics of QRS complex and its derivative. Different number of hidden states are evaluated.

2. Hidden Markov model

An HMM [13] is a statistical model used to characterize signal dynamics as a function of time. A typical HMM is composed of n hidden states and the set of parameters $\lambda = (A, B, \pi)$, where:

(i) $A = a_{ij}$ is the matrix of state transition probability distribution from state i to j.
(ii) $B = b_j(k)$ is the observation symbol probability distribution in state j.
(iii) $\pi = \pi_i$ is the initial state distribution.

A training stage is performed to obtain the set of parameters λ in order to maximize $P(O_{train}|\lambda)$, the probability that an observation sequence $O_{train} = O_1, O_2...O_T$, taken from the analyzed system, is generated by the model.

Figure 1 depicts an example of operation of an HMM, where the observation sequence of length T is emit by a hidden state S_j with probability $b_j(O_k)$. In addition, states transit with probability a_{ij}.

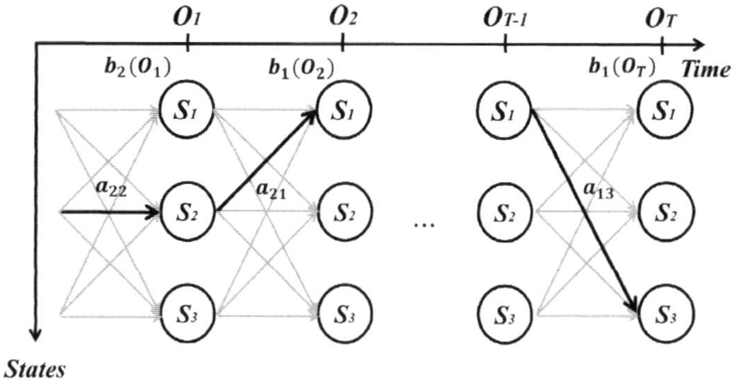

Figure 1. Operation of an HMM.

3. Methods

3.1. ECG dataset

The approach was validated on ECG signals from the MIT-BIH Arrhythmia database [14]. This database contains 48 half-hour excerpts of two-channel ambulatory ECG recordings, sampled at $360Hz$. In addition, annotations of the localization of QRS complexes were provided for each recording.

IOP Conf. Series: Journal of Physics: Conf. Series **1002** (2018) 012009 doi:10.1088/1742-6596/1002/1/012009

3.2. QRS complex detector based on HMM

The ECG signal was filtered with a pass-band $(5 - 15Hz)$ 3rd order Butterworth filter. With the purpose to develop automatic detectors of QRS complexes based on HMM, a training stage exploding the properties of the QRS complex such as the amplitude, $x(t)$, and the rate of change of amplitude, $x'(t) = \frac{dx}{dt}$ was performed. These observations were extracted using an observation window of $T = 55$ samples $(152.8ms)$ centered into each QRS complex annotation provided. In each case, from 40 ECG recordings, 50 QRS complexes were randomly taken from each recording. The resulting HMM, λ_x and $\lambda_{x'}$, trained from $x(t)$ and $x'(t)$, respectively, were analyzed for $2 \leq n \leq 4$.

To begin the detection stage, it was estimated $P(O_{test}|\lambda)$, the probability that an observation sequence, $O_{test} = O_1 O_2 ... O_T$, for $T = 55$ samples $(152.8ms)$, is generated by the model, where O_{test} is extracted from the signals $x(t)$ or $x'(t)$ and is slid along the signal with an overlapping of 98%, and λ can be λ_x or $\lambda_{x'}$. Given the overlapping of the sliding window, we performed the test on the first 14 min of the ECG signal. The $P(O_{test}|\lambda)$ and a fixed threshold β were compared to determine the binary measure of the detector. As λ_x and $\lambda_{x'}$ were adjusted to maximize the probability to obtain a sequence $x(t)$ and $x'(t)$, respectively, then it was stated that if $logP(O_{test}|\lambda) \geq \beta$ a detection of a QRS complex was achieved, where $logP(O_{test}|\lambda)$ is the log-likelihood of observations. A 6-fold cross-validation was implemented to avoid overfitting.

Figure 2 shows the different phases of the detection approach. First, the models are trained in the training phase in order to characterize the dynamics (x_{QRS} or x'_{QRS}). Once the models are trained, they are used to compare $\log P(O_1|\lambda)$ with a threshold on a detection phase. In this phase, we use a sliding window to get the observation sequence along the ECG signal. Finally, annotations and detections are compared in order to obtain the detection performance.

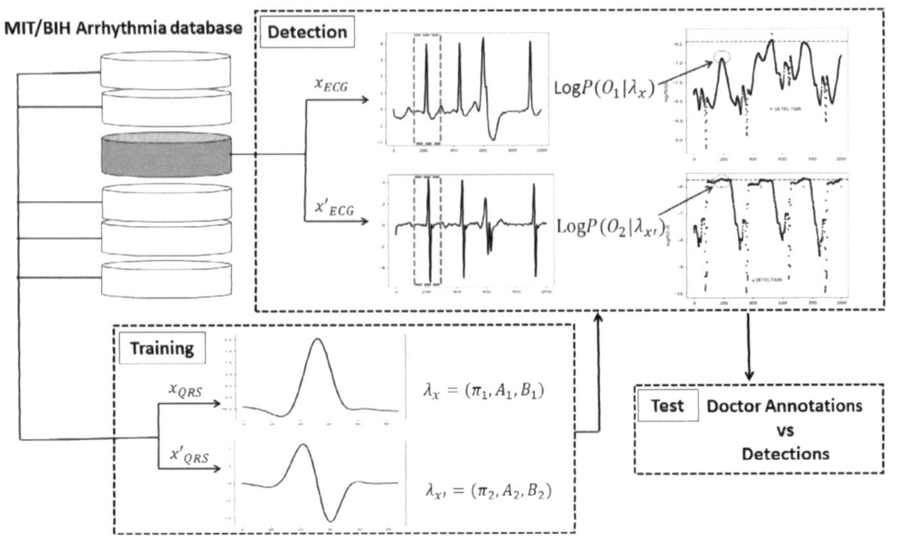

Figure 2. Description of the QRS complex detector based on HMM.

3.3. Performance evaluation

The statistical indicators of each detector were estimated contrasting the measurements of the implemented detector with the professional annotations, allowing to quantify true positives (TP), true negatives (TN), false negatives (FN) and false positives (FP). Then, the performance of each HMM detector was evaluated using sensitivity, $Sen = \frac{TP}{TP+FN}$, specificity, $Spc = \frac{TN}{FP+TN}$ and harmonic mean of sensitivity and specificity, $F_1 = \frac{2TP}{2TP+FP+FN}$. In each case, these indicators were established for different β, with $-10.4 \leq \beta \leq -5.5$.

An optimal threshold, $\beta_{optimal}$, was obtained from the analysis of the receiver operating characteristic (ROC) curve, Sen vs $1 - Spc$. In particular, the distance from a point of the ROC curve to the perfect detection, $\ell = \sqrt{(1 - Sen)^2 + (1 - Spc)^2}$ and the area under the ROC curve (AUC), were quantified.

4. Results

Table 1 shows the results of the mean detection performance for each model, λ_x and $\lambda_{x'}$. In bold the best results found. Despite the AUC was below 0.5, the lowest AUC was 0.387 for $\lambda_{x'}$. The best optimal thresholds for both detector are around -7, these thresholds provide an $F1 = 0.309$ for λ_x and $F1 = 0.293$ for $\lambda_{x'}$. These results show that the QRS complex detection based on HMM using univariate observations, coming either from the ECG signal or its derivative, shows a lower detection performance compared with the state of the art.

Table 1. Mean detection performance for each model for different values of n.

HMM	n	$\beta_{optimal}$	AUC	Sen	Spc	$F1$
	2	−7.5	0.455	0.668	0.360	0.309
λ_x	3	−7.3	0.467	0.612	0.421	0.304
	4	−7.0	0.449	0.547	0.411	0.274
	2	−7.4	0.387	0.765	0.152	0.289
$\lambda_{x'}$	3	−7.2	0.410	0.768	0.166	0.293
	4	−7.1	0.390	0.765	0.171	0.293

5. Discussion

The highest values of AUC and $F1$ score were obtained for the use of the ECG signal instead of its derivate. This can be possible due to the fact that the derivate of the signal could have enhanced the dynamics of P and T waves and bring them closer to the QRS complex dynamics, leading to an increase of FP.

Using both signals, HMM based detectors show a higher number of FP compared with FN (sensitivities were greater than specificities). A reason could be that only QRS complex dynamics were characterized by the model (P and T waves were not taken into account), thus P and T waves could have been confused by QRS complex leading to higher log-likelihood of observations.

The length of the ECG recording being tested is approximately half of the total length, which is a limitation of this paper. This decision was taken given the amount of time involved in the test phase. Future works are directed to the reduction of the overlapping of the sliding window, the characterization of multivariate observations by the HMM, the inclusion of a refractory period to reduce the number of FP, the use of two models to characterize two distinct dynamics, and the use a relative threshold instead of a fixed one to compare the log-likelihood of the observation sequence given the model.

6. Conclusion

Using univariate observations coming from the ECG signal, an HMM was considered to characterize the dynamics of the QRS complex, either from the ECG signal or from its derivative. Our QRS complex detector is based on the comparison of the log-likelihood of an observation sequence given an HMM with a fixed threshold; the best detection performance was $Sen = 0.668$, $Spc = 0.360$ and $F1 = 0.309$ using 2 states and the ECG signal.

References

[1] Sörnmo L and Laguna P 2005 *Bioelectrical signal processing in cardiac and neurological applications* (Amsterdam: Elsevier)
[2] Altuve M, Carrault G, Beuchée A, Pladys P and Hernández A I 2011 *2011 Annual Int. Conf. IEEE Eng. in Med. and Bio. Soc.* p 4374
[3] Cesari M, Mehlsen J, Mehlsen A and Sorensen H 2017 *IEEE J. of Trans. Eng. Health and Med.* **5** 1
[4] Arbateni K and Bennia A 2014 *Neurocomputing* **145** 438
[5] Savalia S, Acosta E and Emamian V 2017 *J. Bio. and Med.* **5** 64
[6] Poli R, Cagnoni S and Valli G 1995 *IEEE Trans. Biomed. Engin.* **42(11)** 1137
[7] Yazdani S and Vesin J M 2016 *Dig. Signal Proc.* **56** 100
[8] Hughes N P, Tarassenko L and Robert S J 2004 *Advances in Neural Information Processing Systems 16* ed Thrun S, Saul L K and Schölkopf B (MIT Press) p 611
[9] Akhbari M, Shamsollahi M B, Sayadi O, Armoundas A A and Jutten C 2016 *Comp. Bio. and Med.* **79** 21
[10] Köhler B, Hennig C and Orglmeister R 2002 *IEEE Eng. Med. and Bio. Mag.* **21** 42
[11] Idoia B and IÃśaki R 2017 *Biomed. Signal Proc. and Control* **34** 166
[12] Elgendi M, Eskofier B, Dokos S and Abbott D 2014 *Plos One* **9(1)** 1
[13] Rabiner L and Juang B 1986 *IEEE ASSP Mag.* **3(1)** 4
[14] Goldberger A *et al.* 2000 *Circulation* **101(23)** e215

Comparative analysis of seismic persistence of Hindu Kush nests (Afghanistan) and Los Santos (Colombia) using fractal dimension

D A Prada[1], M P Sanabria[1], A F Torres[1], M A Álvarez[2] and J Gómez[3]
[1] Universidad Pontificia Bolivariana, Bucaramanga, Colombia
[2] Universidad de Puerto Rico, Mayagüez, Puerto Rico
[3] Universidad Industrial de Santander, Bucaramanga, Colombia

E-mail: duwamg.prada@upb.edu.co

Abstract. The study of persistence in time series in seismic events in two of the most important nets such as Hindu Kush in Afghanistan and Los Santos Santander in Colombia generate great interest due to its high presence of telluric activity. The data were taken from the global seismological network. Using the Jarque-Bera test the presence of gaussian distribution was analyzed, and because the distribution in the series was asymmetric, without presence of mesocurtisity, the Hurst coefficient was calculated using the rescaled range method, with which it was found the fractal dimension associated to these time series and under what is possible to determine the persistence, antipersistence and volatility in these phenomena.

1. Introduction

Fractal geometry as a branch of mathematics allows research on self-similarity and dynamic systems. This branch uses notions of dimension and orbits in various disciplines such as geology, geophysics, medicine, engineering, biology, among others. Earthquakes are geology events, and the complex phenomenology exhibited by them is correlated to the deformation and sudden rupture of parts of the earth's crust because of the external forces acting from plate tectonic motions [1]. Some geophysical processes are characterized by self-similarity correlations because the dynamics are based on the interaction of many components over a wide range of time or space scales [2].

The study of seismic events is important, not only for geologists and geophysicists, but the community in general whom are affected by the natural phenomena. There are places on the planet where such events occur more frequently than others. One area of greatest tectonic activity is Pamir Hindu Kush, in the Afghanistan region [3]. This nest is located at the western syntax of the Himalaya, within the broad deformation belt, area formed by the collision of the Indian and Eurasian lithospheric plates. It is located approximately between the latitudinal boundaries 35-398N and longitudinal boundaries 69-758E [4]. Another geographic region considered as a seismic nest is found in the municipality of Los Santos, situated at $30km$ from the capital of the department of Santander in Colombia. The depth of the seismic events that occur in this zone is in a range of $150 - 170km$ in average. And, according to the National Seismological Network of Colombia RSNC, 50% of the telluric movements recorded in Colombia occur in this municipality; however due, to the depth of the hypocenter are categorized as a low-risk event.

In the period between June 1, 1993, and February 17, 2016, 2502 seismic events were presented in the Hindu Kush zone and Los Santos 3964 seismic events, as recorded by the seismological network world. Each point shows evidence of a telluric event in the area; the color determines depth: violet $0-33km$, blue $33-70km$, green $70-150km$, yellow $150-300km$, orange $300-500km$, and red of $500-800km$.The size of the circle determines its magnitude with respect to the Richter scale [1].

2. Mathematical method

Fractal theory applied to objects allow to construct the notion of self-similarity, dynamic systems, chaos, orbits, and dimension, both topological and Hausdorff, a set of fractal properties that have become useful tools in different disciplines such as mathematics, science, art, etc. [5-7]. Besides the self-similarity, the fractal objects have an idea out of the common; the fractal dimension. The dimension that is assigned by convention to certain geometrical and physical objects is associated with a number infinite variable, for example, a cube is assigned a triple defined directly by the thickness, width, and height thereof, and then the size of this object is three. This type of dimension is known as the topological dimension.

The fractal dimension, as properly its name suggests, is a fractional dimension and is determined by a rational number. Long-range power-law correlations are traditionally measured by a scaling parameter or fractal dimension (D). If the time series is self-similar and self-affine, the parameter D is related to the Hurst exponent (H) through the expression $D = 2 - H$ [8,9]. Thus, the Hurst exponent is a measure of the long-range correlation in time series data and allows to distinguish the persistence (correlation), anti-persistence (anti-correlation) or randomness of the data [10]. The original estimation of the Hurst exponent was first performed in hydrology by Harold Edwin Hurst in 1951 [11], by introducing an empirical relationship called the Rescaled-Range (R/S). Posteriorly, this relationship became the start point to establish the Classical R/S (CR/S) method developed by Mandelbrot and Wallis into the context of the fractal geometry [10,12,13].

Although the CR/S is one of the most popular methods to calculate the Hurst exponent, it has shown some serious limitations to study long-range correlation when the time series is not large enough [13,14]. In the series of seismic events it is necessary to use the modified method of calculating the ratio between the rescaled range and the standard deviation, due to the number of data about CR/S might lead to an overestimation of the Hurst exponent, especially if seismic time series are constructed with less than 5000 data points as happened in this work. On the other hand, a study previously performed using data of Southern México [15] has detected the existence of cycles in the interevent interval sequence and this fact has been considered as a sign of trends in the associated seismic time series.

To apply this method, it is necessary to ensure that the data does not present a Gaussian distribution or that the data are not normalized. The study of these phenomena in front of their distribution may present asymmetry and homoscedasticity, due to their leptokurtic, mesokurtic or platykurtic behavior. In order to determine if the data series presents Gaussian distribution, the Jarque-Bera test was applied. The Jarque-Bera test [16] is performed to corroborate that the time series is not linear, that is to say, that it presents fractalness. For the data analyzed, the asymmetry value was 1.00 and -0.11 for kurtosis. Then the following hypotheses are posed: H0 (null hypothesis): Mean=0; Variance=1, i.e., the normal distribution; H1: Mean≠0; Variance≠1, i.e., the distribution is not normal. To find the value of the Jarque-Bera test (JB), the following equation:

$$JB = n\left[\frac{S^2}{6} + \frac{(K-3)^2}{24}\right] \tag{1}$$

Where S represents the asymmetry and K the kurtosis. The Jarque-Bera coefficient is greater than 6, so the series does not show normality.

Specifically, a value $0 < H < 0.5$ corresponds to anti-correlated data (anti-persistence behaviour), a value $0.5 < H < 1$corresponds to correlated data (persistence behaviour) and the value $H = 0.5$ corresponds to random data (uncorrelated behaviour) [10]. The definition of H can be extended to values

larger than 1 [12]. In this way, the case $H = 1.5$ corresponds to Brownian motion, the case $H = 2$ corresponds to brown noise and the case $H > 2$ corresponds to black noise [17].
Shown below CR/S method the time series described in [18] under consideration $X: \{x_i\}$ is composed by N values. The full-time series is divided into windows of size M. The number of windows is defined by $s = N/M$ and therefore there are s windows of data Y_j, with $j = 1,2, \dots, s$.
Defining the vector $k = (j-1)M + 1, (j-1)M + 2, (j-1)M + 3, \dots, (j-1)M + M$, the average over each window is calculated as:

$$\bar{Y}_j = \frac{1}{M} \sum_k x_k \tag{2}$$

The profile or sequence of partial summations $Z_j: \{z_n\}$, with $n = 1,2, \dots, M$ is defined as the cumulative summation minus the average of the corresponding window

$$z_n = \sum_k^n \{x_k - \bar{y}_j\} \tag{3}$$

The range R_j of the window is defined as the maximum minus the minimum data point of the profile

$$R_j = max\{Z_j\} - min\{Z_j\} \tag{4}$$

The standard deviation of each window σ_j is given as:

$$\sigma_j = \left[\frac{1}{M} \sum_k (x_k - \bar{y}_j)^2\right]^{\frac{1}{2}} \tag{5}$$

The rescaled range is described by the quantity $(\mathrm{R/S})_\mathrm{M}$, which is defined as

$$(\mathrm{R/S})_\mathrm{M} = \mathrm{mean}\{R_j/\sigma_j\}. \tag{6}$$

For the case in which a stochastic process associated to the data sequence under study is rescaled over a certain domain $M \in \{M_{min}, M_{max}\}$, the R/S statistics follows the power law

$$(\mathrm{R/S})_\mathrm{M} = aM^H \tag{7}$$

Herein, a is a constant and H is the Hurst exponent which represents a fractal measure of the long-range correlations in the analysed data. However, it is necessary to use the improved method of the rescaling range, in which autocovariance and its autocorrelation are used. The equation (5) of the standard deviation changes for this calculation by the equation:

$$\sigma_j(q) = \sqrt{\sigma_j{}^2 + 2\sum_{k=1}^q \omega_k(q)\, \gamma_k}, \omega_k(q) = 1 - \frac{k}{q+1} \tag{8}$$

Where $\sigma_j{}^2$ is the sample variance and γ_k the sample autocovariance. The parameter q represents the lags in the weighted autocovariances.

3. Results and discussion

The time series of seismic events occurred between June 1, 1993, and February 17, 2016, in the Hindu Kush Afghanistan and Los Santos Colombia zones, presented 2502 and 3964 events respectively, as recorded by the global seismological network. Each group of data was divided into as many groups as divisors greater than 100. Table 1 shows the group of the first 139 data recorded in the Hindu Kush area from June 6, 1993, to August 27, 1996.

Table 1. Data of the standard deviation, data number, and rescaled range.

Place	Standard deviation	Number of data	Rescaled range	Max	Min	Average
Hindu Kush	0.630957685	139	13.22446	13.22446	3.934E-13	4.097122
Los Santos	0.423422389	283	20.18763	20.18763	3.117E-13	4.139929

With each of the series the same procedure is performed, in Table 2 and Table 3 the necessary data for the calculation of the Hurts coefficient for this time series are shown, in addition, we compare the natural logarithm of the data of each of the subgroups generated versus the natural logarithm of the ratio between the rescaled range and standard deviation.

Table 2. Related data from the four groups depending on the range and standard deviation. Calculation of the natural logarithm of the data number and the natural logarithm of the quotient between the rescaled range and the standard deviation. Hindu Kush.

Number of data	Rescaled range	Standard deviation	ln(num)	ln(R/S)
139	13.22446043	0.630957685	4.934473933	3.042584656
278	25.49064748	0.596857382	5.627621114	3.754388704
417	37.37266187	0.568646170	6.033086222	4.185436354
834	49.42829736	0.580429930	6.726233402	4.444509273
1251	99.99904077	0.598812745	7.131698510	5.117966936
2502	383.97274180	0.723505287	7.824845691	6.274218991

Table 3. Related data from the four groups depending on the range and standard deviation. Calculation of the natural logarithm of the data number and the natural logarithm of the quotient between the rescaled range and the standard deviation. Los Santos.

Number of data	Rescaled range	Standard deviation	ln(num)	ln(R/S)
283	20.18763251	0.423422389	5.645446898	3.864455207
566	66.10477032	0.506678775	6.338594078	4.871118969
1983	316.59671880	0.656226596	7.592366129	6.178877913
3962	460.88414940	0.644585530	8.284504227	6.572294466

With the results shown in Table 2 and Table 3, the data of the last two columns are taken, the column of the natural logarithm of the data number and the column of the natural logarithm of the rescaled range over the standard deviation of each subseries. With the data of these two columns, a linear regression is elaborated whose abscissa is the natural logarithm of the data number, and the ordinate corresponds to the natural logarithm of the rescaled range over the standard deviation. Figure 1(a) shows the graph of the linear regression and the scatter data. The value of the slope of said linear equation is 1.0427 with a correlation coefficient R^2 of 0.9566, which represents the associated fractal dimension of the data corresponding to the Hindu Kush seismic events. Figure 1(b) shows the graph of the linear regression and the dispersion data corresponding to the data of the seismic events in Los Santos, the value of the slope of said linear equation is 1.092, which represents the fractal dimension associated to these dates. The correlation coefficient R^2 is 0.9797.

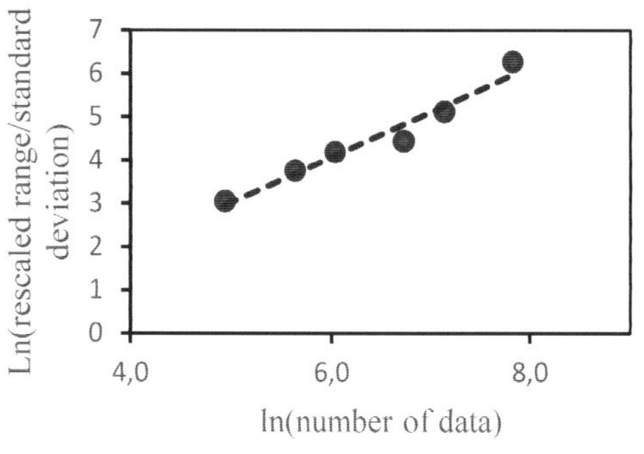

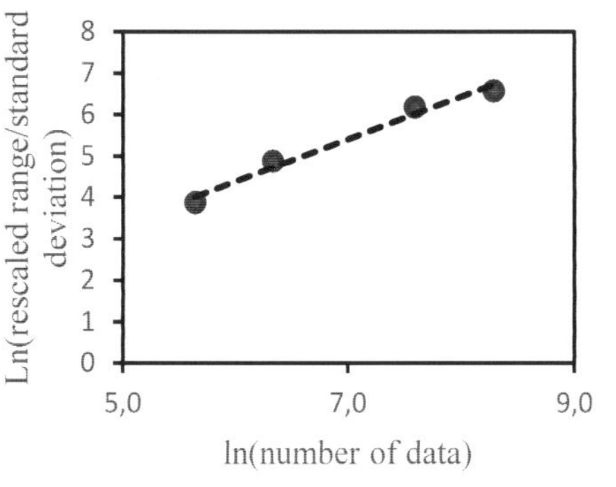

Figure 1. (a) Logarithmic linear regression of the data supplied in Table 2. (b) Logarithmic linear regression of the data supplied in Table 3.

4. Conclusions

From the data recorded by the world seismological network between June 1, 1993, and February 17, 2016, in the areas of Hindu Kush Afghanistan and Los Santos Colombia, which presented 2502 and 3964 events respectively, the Hurts coefficient was $H = 1.0427 > 0.5$ for Hindu Kush with a fractal dimension of $D = 0.9573$ and for Los Santos the coefficient was $H = 1.0292 > 0.5$ with a fractal dimension of $D = 0.9768$, which indicates that the time series presents low fractal dimension and for this reason the series present persistence associated with a volatility of 47.86% for Hindu Kush and a volatility of 48.54% for Los Santos. These series did not present Gaussian distribution because the test of Jarque-Bera was greater of 6 for each one.

References

[1] Michas G, Vallianatos F and Sammonds P 2013 *Non. Proc. Geo.* **20** 713–724
[2] Ashkenazy Y, Havlin S, Ivanov P, Peng C, Schulte-Frohlinde V and Stanley H E 2003 *Phys. A* **323** 19–41
[3] Bai L and Zhang T 2015 *Tecto.* **638** 177–184
[4] Sarkar I and Sanyal S 2003 *J. Asi. Ear. Sci.* **23** 449–459
[5] Mandelbrot B 2006 *Los objetos fractales. Forma, azar y dimensión* (España: Metatemas)
[6] Falconer K 2006 *Fractal geometry: Mathematical foundations and applications* (India: Wiley)
[7] Barnsley M 2012 *Fractals Everywhere* (United States: Dover Publications)
[8] Feder J 1988 *Fractals* (New York: Plenum Press)
[9] Chen C, Lee Y and Chang Y 2008 *Phys. A* **387** 4643–4648
[10] Mandelbrot B and Wallis J 1968 *Water Resour* Res. **4** 909–918
[11] Hurst H E 1951 Long-term storage capacity of reservoirs *Trans. Am. Soc. Civ. Eng.* **116(1)** 770–799
[12] Mandelbrot B and Wallis J 1969 *Wat. Res. Res.* **5** 321–340
[13] Mandelbrot B and Wallis J 1969 *Wat. Res. Res.* **5** 967–988
[14] Sánchez M, Trinidad J and García J 2008 *Phys. A* **387** 5543–5551
[15] Álvarez J, Echeverría J C, Ortiz A and Hernández E 2012 *Geo.* **54** 1–12
[16] Andrew W 1991 *Econ.* **59** 1279-1313
[17] Jimenez A 2011 *Phys. A* **390** 2146–2154
[18] Martin L, Aranda N and Quimbay C 2015 *Phys. A* **421** 124–133
[19] Gujaratí D and Porter D 2010 *Econometría* (Mexico: Mc Graw Hill)

IOP Conf. Series: Journal of Physics: Conf. Series **1002** (2018) 012011 doi:10.1088/1742-6596/1002/1/012011

Analysis of persistence in fluctuation of the Cauca river through the Hurst coefficient

D A Prada[1], M P Sanabria[1], A F Torres[1], A Acevedo[1] and J Gómez[2]
[1] Universidad Pontificia Bolivariana, Bucaramanga, Colombia
[2] Universidad Industrial de Santander, Bucaramanga, Colombia

E-mail: duwamg.prada@upb.edu.co

Abstract. Study the continuous changes in the fluctuations in the levels of watersheds, it is of great importance because it allows you to adjust predictions about behaviors that can lead to floods or droughts. The Cauca River is one of the most important rivers in Colombia due to its $1350km$ long, its drainage area of $59.074km^2$ which represents 5% of the national territory. The Government entity Cormagdalena records daily levels of the Cauca River to the height of *La Mojana* in the rods. From these data, we developed a series of time on which normal test were applied to calculate the coefficient of Hurst and the fractal dimension to determine the persistence associated with this behavior.

1. Introduction

The basins are highly complex systems, their formation process is linked to factors as variables in time and space, such as climate, geology, vegetation etc. However, its evolution is subject to the combination of these forces with the climatic factors.

The measures the water level in a river contribute to understanding the dynamics of the download at a specific point in the basin. With this type of information, it is possible to predict and consider future flood events. The flow in the rivers also depends directly on the speed that it has, if the speed is small then the flow is regular, however, when you pass a stone, the river surrounding it, and if your speed increases becomes irregular, creating swirls and eddies in these eddies. The hydrodynamics describes these movements of initial form through the laws of Newton for fluids, in this way was found evidence of a non-linear equation, the equation of the Navier-Stokes Equations [1]. In the particular case of drainage networks the famous British hydrologist Harold Edwin Hurst (1880-1978), study the fluctuations in the level of the River Nile, in a large part of its extension, to be able to project the capacities of the reservations and take precautionary measures in times of drought [2].

2. Cauca's river

The Cauca river is one of the most important water resources in Colombia. It has a length of $1350km$, with a basin area of approximately $63300km^2$. It goes across the country from south to north through nine departments and a number of cities and towns without appropriate wastewater treatment plants [3].

The Cauca river is born in the Colombian Massif between the Western and Central Cordilleras of the Andes with a drainage area of $59.074km^2$ that represents the 5% of the national territory. In the course of its first $578km$ (48% of the total) conforms the basin in its top-half the Colombian southwest [4]. The periods where there is rain in the months of March-April-May and September-October-November, and periods of low precipitation are December-January-February and June-July-August [5]. The region has an average annual rainfall of $1597mm$ and $21.3°C$ average air temperature [4]. The Government entity Cormagdalena records every day of the Cauca river levels to the height of the sector of *La Mojana*, place that includes the municipalities of Sucre, Guaranda and Majagual in the southern part of Department of Sucre, as seen in the Figure 1.

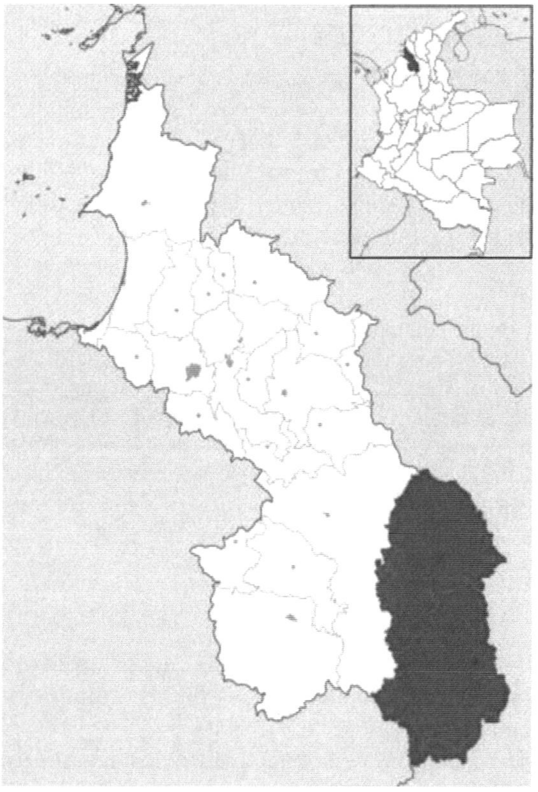

Figure 1. *La Mojana* region [6].

Study the continuous changes in the fluctuations in the levels of watersheds, it is of great importance because it allows you to adjust predictions about behaviors that can lead to floods or droughts. The first to recognize the characteristics of invariance of the drainage network to establish the laws of scale was Horton, and denoted as: Relationship of fork (R_B), a ratio of length (R_L) and (R_A) [7]. $R_L R_A$Natural objects that have a complex spatial structure can be studied as fractals. Fractal geometry to describe irregular patterns and fragmented, which are repeated at different scales, usually in isotropic [8]. Fractal geometry as a branch of mathematics has been allowed to carry out research in relation to the similarity and dynamic systems, using notions of dimension and orbits in various disciplines such as medicine, engineering, psychology, music, biology, among others. In the particular case of drainage networks two fractal dimensions can be calculated, a fractal dimension (d), which describes the sinuosity of a current and the fractal dimension of the entire network (D), related to the characteristics of branching or compactness of the system [9,10]. With the analysis data between the length of the currents and the area of drainage basins, he found that the Mandelbrot fractal dimension of water courses (d) was equal to 1.2, while the fractal dimension of the entire system (D) was equal or close to 2 [11].

3. Mathematical method

The Fractal Geometry is a branch of mathematics, very attractive fractal objects that are possible to construct the notion of self-similarity, dynamic systems, chaos, orbits and dimension, both topological and Hausdorff spaces, which have become useful tools in numbers disciplinary fields such as mathematics, science, art, etc [11-13]. Besides the self-similarity, the fractal objects have an idea out of the common; the fractal dimension. The dimension that is assigned by convention to certain geometrical and physical objects, is associated with a number infinite variable, for example, to cube is assigned to triple defined directly by the thickness, the width and height thereof, and then the size of this object is three. This type of dimension is known as the topological dimension [11,13].

The fractal dimension, as properly its name suggests, is a fractional dimension and is determined by a rational number. Long-range power-law correlations are traditionally measured by a scaling parameter or fractal dimension (D). If the time series is self-similar and self-affine, the parameter d is related to the Hurst exponent (H) through the expression $D = 2 - H$ [14,15]. Thus, the Hurst exponent is a measure of the long-range correlation in time series data and allows to distinguish the persistence (correlation), anti-persistence (anti-correlation) or randomness of the data [16]. The original estimation of the Hurst exponent was first performed in hydrology by Harold Edwin Hurst in 1951 [17], by introducing an empirical relationship called the Rescaled-Range (R/S). Posteriorly, this relationship became the start point to establish the Classical R/S (CR/S) method developed by Mandelbrot and Wallis into the context of the fractal geometry [16,18,19]. Although the CR/S is one of the most popular methods to calculate the Hurst exponent, it has shown some serious limitations to study long-range correlation when the time series is not large enough [19,20].

4. Respect to the test of normality

Because of the behaviour variant that presents the time series of the fluctuating levels of depth of the river Cauca, were used Anderson-Daling Smirnov-Kolmogorov test and to ensure that the data do not present a Gaussian distribution, that is to say that the data are not standardized. The study of these phenomena in front of your distribution may present asymmetry and Homosedasticidad, because your behaviour platicurtico mesocurtico leptocurtico, or for these tests has been raised as a null hypothesis H0: there is a presence of normality (there is no presence of trend in the series) and how alternative hypothesis H1: There is a trend in the time series. In Figures 2 and 3 shows the analysis of normality for the time series using the test mentioned above, under the condition that if the value $\rho <$ significance (0.05 usually) reject the null hypothesis and if the value $\rho >$ significance (0.05 usually) accepts the null hypothesis.

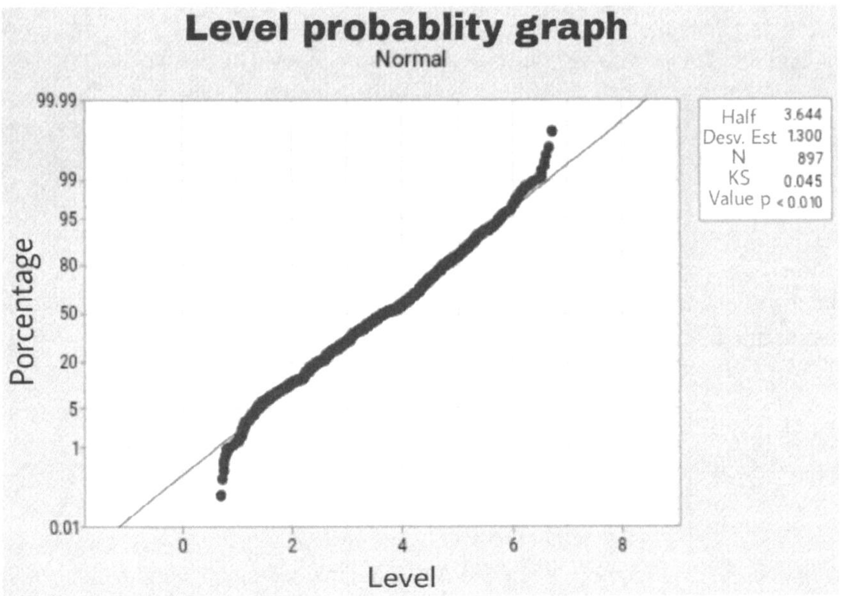

Figure 2. Analysis of normal

With the data obtained in previous tests, we reject the null hypothesis and therefore it can be seen that there is a behaviour of similarity in the time series analysed.

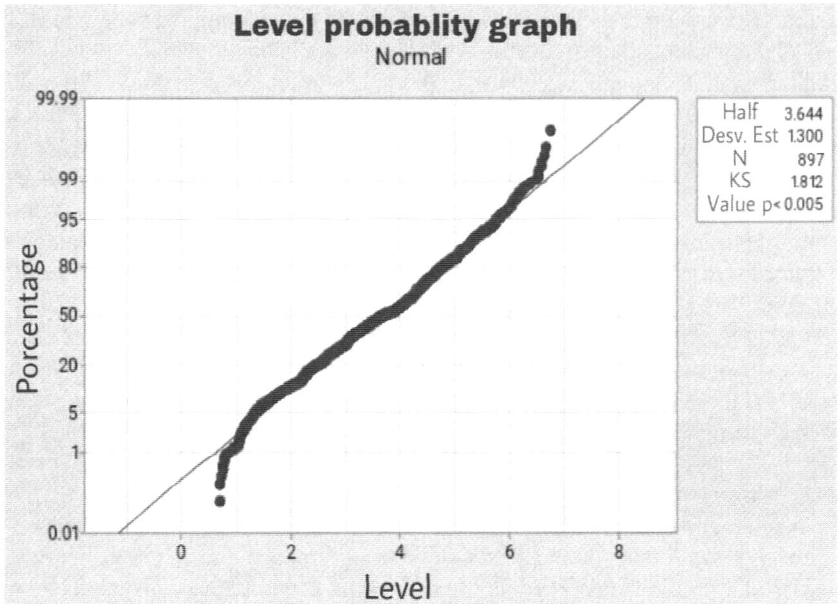

Figure 3. Smirnov-Kolmogorov Test Anderson-Daling Test.

For the analysis of fractal dimension of the time series, specifically, to value $0 < H < 0.5$ corresponds to anti-eje correlated released (anti-persistence behaviour), a value $0.5 < H < 1$ corresponds to correlated axis released (persistence behaviour and the value $H = 0.5$ corresponds to random data (uncorrelated behaviour [16]. The definition of H can be extended to values larger than 1 [18]. In this way, the case $H = 1.5$ corresponds to Brownian motion, the case H=2 corresponds to brown noise and the case $H > 2$ corresponds to black noise [21].

The following is CR/S method the time series described in [22] under consideration $X{:}\{x_i\}$ is composed by N values. The full-time series is divided into windows of size M. The number of windows is defined by $s \equiv N/M$ and therefore there are s windows of data Y_j, with $j = 1,2,\dots,s$. Defining the vector $k = (\mathrm{j}-1)\mathrm{M}+1, (\mathrm{j}-1)\mathrm{M}+2, (\mathrm{j}-1)\mathrm{M}+3,\dots,(\mathrm{j}-1)\mathrm{M}+\mathrm{M}$, the average over each window is calculated as:

$$\bar{Y_j} = \frac{1}{M}\sum_k x_k \quad (2)$$

The profile or sequence of partial summations $Z_j{:}\{z_n\}$, with $n = 1,2\dots M$, is defined as the cumulative summation minus the average of the corresponding window

$$z_n = \sum_k^n\{x_k - \bar{y_j}\} \quad (3)$$

The range R_j of the window is defined as the maximum minus the minimum data point of the profile

$$R_j \equiv max\{Z_j\} - \min\{Z_j\} \quad (4)$$

The standard deviation of each window σ_j is given as

$$\sigma_j = \left[\frac{1}{M}\sum_k (x_k - \bar{y_j})^2\right]^{\frac{1}{2}} \quad (5)$$

The rescaled range is described by the quantity $(R/S)_M$, which is defined as

$$(\mathrm{R/S})_{\mathrm{M}} \equiv \text{mean } (R_j/\sigma_j) \tag{6}$$

For the case in which a stochastic process associated to the data sequence under study is rescaled over a certain domain $M \in \{M_{min}, M_{max}\}$, the R/S statistics follows the power law

$$(\mathrm{R/S})_{\mathrm{M}} = \mathrm{a}M^H \tag{7}$$

Herein, is a constant and H is the Hurst exponent which represents a fractal measure of long-range correlations in the analysed released.

5. Results and discussion

The Government entity Cormagdalena recorded 897 data from the Rio Cauca levels in the *La Mojana* sector on the date between September 29, 2011 and September 1, 2017. With this amount of data 6 subgroups were built, the first sub-group with 64 data falling between 29 September 2011 until 16 February 2014, the second subgroup with 112 data falling between 29 September and 23 May 2012, the third subgroup with 128 data with dates of 29 September until 8 August 2012, the fourth subgroup with 224 data with dates of 29 September until 3 November 2013, the Fifth 488 subgroup data and dates between 29 September and 19 February 2015, and the last Subgroup with the totality of the data. This division is accumulated to observe the self-similarity in the behaviour of the data as you increase the amount of the same. The number of subgroups that have been generated for this series is related to the dividers on the total number of data, i.e. with the number 897. This is possible because no matter the scale at which the analysis is carried out, the behaviour must be the same based on the fractality of the series. In each of this series is calculated as the standard deviation, the average and range- scaled by the difference of the maximum and minimum value of the cumulative sum of deviations. Table 1 describes the data obtained in each subgroup.

Table 1. Data of the standard deviation, released number and rescaled range.

Subgroup	Call forwarding stand	Number of released	Rank rescaled	Max	Min	Average
1	0.59	64	14.96	1.77	-13.20	5.26
2	1.03	112	42.14	42.14	0	4.60
3	0.98	128	46.05	42.47	-3.57	4.60
4	0.93	224	53.02	34.54	-18.49	4.72
5	1.26	488	190.79	170.42	-20.36	3.96
6	1.30	897	289.76	277.23	-12.53	3.64

In the Table 2 shows the data needed for the calculation of the coefficient of hurts for this time-series, as well as compare the natural logarithm of the data of each of the subgroups generated versus the natural logarithm of the ratio between the rescaled range and standard deviation. With this data we make a new graph in which we compare the natural logarithm of the data of each of the subgroups generated versus the natural logarithm of the ratio between the rescaled range and standard deviation, these data are related in Figure 4.

Table 2. Related data from the four groups depending on the range and Standard deviation. Calculation of the natural logarithm of the data number and the natural logarithm of the quotient between the rescaled range and the standard deviation.

Subgroup	Number data	Rescaled range	Standard Deviation	Ln(Num)	Ln(R/S)
1	64	14.96	0.58793287	4.158883083	3.236835771
2	112	42.14	1.03	4.718498871	3.712918796
3	128	46.05	0.98113714	4.852030264	3.848672389
4	224	53.02	0.93105402	5.411646052	4.042147592
5	488	190.79	1.25977654	6.190315406	5.020233087
6	897	289.76	1.30115687	6.799055862	5.405807777

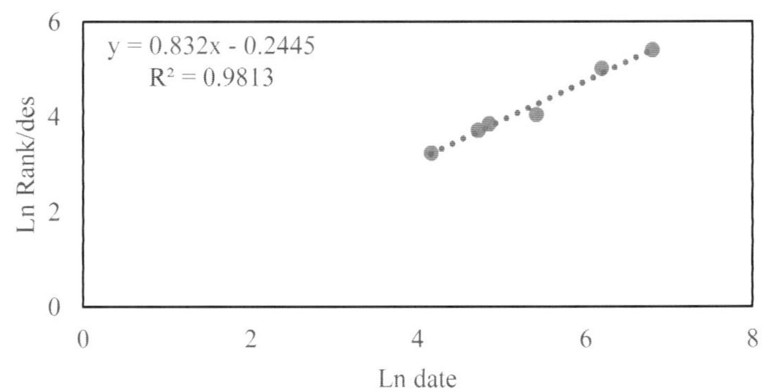

Figure 2. Logarithmic linear regression of the data supplied in Table 2.

6. Conclusions

Of the 897 data recorded by the governmental entity Cormagdalena between el 29 September 2009 June 1993 and 1 September 2017, in the area of *La Mojana*-Las Rods with respect to the levels of fluctuation of the Magdalena River, the coefficient of Hurts associated was of $H = 0.832 > 0.5$, which induces that the fractal dimension is $D = 1.168$, according to [14,15], which indicates that the time series of low fractal dimension and for that reason these series have persistence associated with a volatility of 58.4%. In addition, it was found that this basin meets the above mentioned by Mandelbrot Set with respect to the fractal dimension of the basins with these characteristics must be close to the value of 1.2. These series were not gaussian distribution due to the fact that the Jarque-Bera test was greater than 6 for each [23].

References

[1] Braun E 2011 *Caos, fractales y cosas raras* (Mexico: Fondo de Cultura Económica) p 92
[2] Rodriguez E 2012 *Lat. Amer. Not. Adm.* **3** 42-50
[3] Ocampo W, Osorio C, Schuhmacher M and Domingo J 2012 *Env. Int.* **52** 17-28
[4] Puerta O, Carvajal Y and Quintero M 2011 *Dyna* **169** 112-120
[5] Mesa O, Poveda G and Carvajal L 1997 *Introducción al clima de Colombia* (Colombia: Universidad Nacional de Colombia)
[6] Escuela Superior de Administración Pública ESAP. Consulted on http://cdim.esap.edu.co/Combosdependientes.asp?PnDepartamentos=70&Pnmuni=70771
[7] Ramirez R 2017 *Min. and Geo.* **33** 279-292
[8] Ramirez R, Rodriguez A and Ordaz A 2017 *Min. and Geo.* **33** pp 36-49
[9] Barbera P and Rosso R 1989 *Wat. Res. Res.* **25** 735-741

[10] Tarboton D 1996 *J. Hyd.* **187** 105-117
[11] Mandelbrot B 2006 *Los objetos fractales; forma, azar y dimensión* (España: Metatemas)
[12] Falconer K 2006 *Fractal geometry. Mathematical foundations and applications, second edition* (India: Wiley)
[13] Barnsley M 2012 *Fractals Everywhere* (United States: Dover Publications)
[14] Erdf J 1988 *Fractals* (New York: Plenum Press)
[15] Chen C, Lee Y and Chang Y 2008 *Phys. A* **387** 4643-4648
[16] Mandelbrot B and Wallis J 1968 *Wat. Res. Res.* **4** 909-918
[17] Hurst H 1951 *Trans. Am. Soc. Civ. Eng.* **116** 770-808
[18] Mandelbrot B and Wallis J 1969 *Wat. Res. Res.***5** 321-340
[19] Mandelbrot B and Wallis J 1969 *Wat. Res. Res.* **5** 967-988
[20] Sanchez M, Trinidad J and García J 2008 *Phys. A* **387** pp 5543 -5551
[21] Jiménez A 2011 *Phys A* **390** 2146-2154
[22] Martin L, Aranda N and Quimbaya C 2015 *Phys. A* **421** 124 -133
[23] Gujarati D and Porter D 2010 *Econometría* (Mexico: Mc Graw Hill)

IOP Conf. Series: Journal of Physics: Conf. Series **1002** (2018) 012012 doi:10.1088/1742-6596/1002/1/012012

About the Law of Sinus in a trigonometry class

D A Prada[1], J Mantilla[1], A Díaz[1], F Páez[1] and J Gómez[2]
[1] Universidad Pontificia Bolivariana, Floridablanca, Colombia
[2] Universidad Industrial de Santander, Bucaramanga, Colombia

E-mail: duwamg.prada@upb.edu.co

Abstract. The law of sine is an equation of great utility in trigonometry. This type of equation has been applied in various contexts as the analysis of the relationship between angle of formation and sheet thickness of aluminum in processes of embossment, also in an instance in the duality of polar spaces of constant curvature. The applications are made obvious in the sense that it is analyzed and continually ponders the theoretical formality. In this article, we show a our own theorem and_corollary which is useful in trigonometry.

1. Introduction

Continuous progress to which we face as a society of knowledge, both in the technological part as academic, demand of us a propositive stance to change. This is possible to achieve provided that reform and transform the mediation of knowledge, as a process of reflection, in which each actor involved is recognized and recognizes the other as a pillar in the construction of appropriate policies to routing processes effectively to society. At this moment we are fortunate because educational institutions tend for the integral development of all the members of the academic community, especially the researchers in training, as the person trained integrally is autonomous in their learning, is able to generate their own criteria based on in the things they observes, looks, listens and reads [1]. On the other hand, although training by competences carries in its DNA the constant search for the development and continuous improvement of people [2], the process of integral formation is continuous, permanent, participative and seeks to harmoniously and coherently develop each and every one of the dimensions of the human being [3] and can not be based solely on the transmission of information or only on the teaching of competencies [4].

One of the spaces in which it is possible to generate improvements in the competences of the students is in the formative research, taking a leading role for this purpose the research seedbeds, which emerge as an alternative in the need of transformation of the pedagogy to the new educational reality taking advantage of the capacity of characteristic astonishment of the young people, impelling them and orienting them so that they develop investigative competences [5], since sometimes it is considered that in the research seedbeds, students must generate and execute directed work proposals to problem-solving [6].

All this must be guaranteed in favor of the development of the ability and capacity of the human being to know how to interpret and appreciate the content of the mind of others in order to understand and learn from others [7], understanding learning as building and restructure models to interpret the information it receives [8] seeking the applicability of it for the development of the community, which is why it is considered of great importance that universities guarantee citizen competence of their students [9].

At the academic level, this type of reflection is inherent both in teachers and students. It is necessary to create strategies that enable innovation in the processes of teaching and learning in the classroom and outside of it, where the topics are not exclusively for the focused subject. Cross-cutting knowledge allows, both teachers and students, to relate the theoretical concepts and the pilot, having as purpose the proposition improvements regarding the context.

Institutional educational projects, middle vocational level, covers subjects of great importance, that initiated the theoretical formalism required to the University level , among them there is Trigonometry, which in its teaching-learning process presents a great variety of conceptions and difficulties associated with the concepts of this area of knowledge, as evidenced by various research carried out in this field in recent decades [10-12]. Reason for which some institutions of higher education, contemplate the inclusion of geometry and trigonometry courses, primarily to strengthen necessary topics in the development of the professionalizing cycle.

Trigonometry is one of the courses of the area of mathematics of wider applicability in physics, engineering, chemistry, aeronautics, among others [13]. In trigonometry, a study is made more formal of the relationship of measures of a triangle, basically between sides and angles. The trigonometric functions because of their journal behavior has been used for modeling seismic waves, sounds, in medicine they have used trigonometry for several studies that involve the implant inclined to patients [14], in the same way, that has been applied to define the relationship between bilateral pedicle screws and vertebral rotation in which vertebral rotation is calculated after a postoperative period [15], trigonometry has also been used to evaluate abduction of lower limbs in newborns [16]. Goniometry is a method commonly used to measure the movement of the joint arch, particularly in relation to range of motion and muscle tone [17-19], similarly, in [20] a method based on trigonometric relations is presented between the distance of the work ports to the optical port and the distance between the optical port and the surgical objective, in order to provide a tool to help the reproducible evaluation of the laparoscopic surgical technique. On the other hand, in the field of engineering, trigonometry has been used to perform various calculations that involve objects not only triangular in shape, in [21-25] the trigonometric leveling method is used to determine the orthometric height of the points located in areas of difficult access, such as mountainous regions. Similarly, trigonometry has been used to determine joint angles in the construction of metamorphic robotic hands with spherical palms [26]. Law of sine and cosine law is used to analyze and find the relationships in triangles that are not rectangles. The law of Sines and law of cosines are of utmost importance in the field of trigonometry as these theorems establish relationships between the three sides and three angles of any triangle [27]. Sine law lays down that the equation is satisfied for a triangle [28,29].

$$\frac{sen(\alpha)}{a} = \frac{sen(\beta)}{b} = \frac{sen(\theta)}{c} \tag{1}$$

Where α, β and θ they are the angles of the triangles and a, b and c he corresponding opposite sides. In our university courses it is, a law applied in contexts in which it is possible to observe or construct a triangle associated data, for example the distances, or the calculation of angles of elevation, among others. There are other applications for this law among which is located as a fundamental contribution to the experimental geometry in various processes to increase the thickness of a sheet. For example in the construction of aluminium sheets, where the particular geometry uses the dimensional properties, metrology, the relationship between the thicknesses and angles with which law of the sinus for embossment can be verified [30]. Figure 1 shows that for a simple conical form, the resulting reduced wall thickness can be determined easily through the lye of the breasts [30]

There are also studies more advanced concerning the use of this law, where arises, advanced as an instance in the duality of polar spaces of constant curvature [31]. This advanced form is represented by:

$$\frac{sen(\alpha)}{S_k(a)} = \frac{sen(\beta)}{S_k(b)} = \frac{sen(\theta)}{S_k(c)} \tag{2}$$

Under the same conditions for the angles and sides, just that defines the generalized sine function:

$$S_k(x) = \begin{cases} \dfrac{\sin(\sqrt{k}x)}{\sqrt{k}} & for\ k > 0 \\ x & for\ k = 0 \\ \dfrac{\sinh(\sqrt{|k|}x)}{\sqrt{|k|}} & for\ k < 0 \end{cases} \tag{3}$$

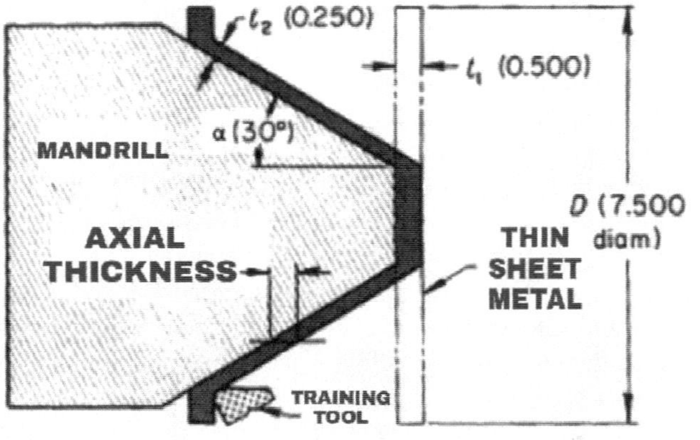

Figure 1. Relationship between angle of training and using sheet metal thickness law of sine in processes of embossment [30].

A basic principle of teaching mathematical methods is to start using examples so students can easily follow the technique [32]. However, it is important that students understand the formality of this type of theorems, in order to enable them to deal with different ways of thinking. With the generation of these spaces, teaching mediation can become the canal's construction of new theories, the use of known laws helps to discover others using basic methods such as algebraic substitution [27]. There are other relationships in trigonometry that are basically intact, such as the law of the tangents and Mollweide formula. The latter is a set of two relationships between the sides and angles of a triangle, and this formula presents property of symmetry similar to the sine and the cosine law, which implies that they are cyclical in nature and thus for a law to another simply turn the order of sides and angles [33].

2. Mathematical method

Sine law also introduces an ambiguity, since there may be up to two satisfactory answers depending on the data supplied initially. The basic condition is to show three data among which involved an angle, its opposite side, and other any data, that may be another angle or other side. Case to case see the possibilities. In the first case, in which the third Datum is one angle more, immediately have the courage of all the Interior of the triangle angles due to the fact that given two of them, the sum of the three is equal to 180°, i.e. that the third angle will be equal to the difference between 180° and the sum of the two given angles. Under this condition, there are two possibilities; there is no response, case in which it is noted that the bosom of one of the angles has value less than -1 or greater than 1; the other condition is that there is a single response that is satisfied. Now if data given initially offered, an angle, its opposite side, and another side, then there are three conditions, the first is that there is no response or that there is only one like the previous case. However, it is possible that there is another answer. Knowing the angle α, and the sides a and b, we can find the value of the angle β using equation (1), and with these data find the angle missing by the sum of the internal angles 180°. If in addition, the supplementary angle β, call it β', coupled with the given angle α is initially less than 180°, implies that there is another triangle satisfying the initial conditions. Take for example the case in which you want to know the missing data if they $\beta = 50°$, $b = 5$, and $c = 6$. Of the law of Sines, we have that

$$\frac{sen(50°)}{5} = \frac{sen(\theta)}{6} \tag{4}$$

It is that $\theta = 66.82°$. Then it follows easily that $\alpha = 63.18°$ and $a = 5.83$, but as seen, must be $\theta' = 113.18°$ and adding with the initially given angle is fulfilled that $\beta + \theta' = 163.18° < 180°$, for which there is a second solution satisfying the initial conditions where $\alpha = 16.82°$ and $a = 1.89$.

A question is to know if given three data that satisfy the condition for the use of the sine law, presents or not two possible solutions. A continuous reflection of this question was the following theorem that does not give a solution to the question above but shows a simple relationship between the Interior and exterior angles of a triangle oblique.

Theorem. The triangle ABC is oblique, with interior angles α, β, and θ, and sides a, b, and c respectively. They are also α', β', and θ' respective external angles of the triangle. Given an interior angle, the value of this will always be less than any different to their supplementary exterior angles measure.

Demonstration. Without loss of generality, is the interior angle α. We need to show that $\alpha < \beta'$ and $\alpha < \theta'$. Note that $\alpha + \beta + \theta = 180°$, in the same way as β' is the supplement of β, we have to $\beta' + \beta = 180°$. Then we have that $\alpha + \beta + \theta = \beta' + \beta$, then $\alpha + \theta = \beta'$ and given that $\theta > 0$, has to be $\alpha < \beta'$. The case $\alpha<\theta'$ show is done in a similar way. Of the previous theorem can be inferred in the following immediate corollary.

Corollary. The triangle ABC is oblique, with interior angles α, β and θ, and sides a, b, and c respectively. They are also α', β' and θ' respective external angles of the triangle. If two interior angles are equal, then under the sine law, there will be a single triangle maximum that meets the basic conditions given.

3. Results

Sine law has been used for various physical applications such as training and thickness of chapas of aluminum, the duality of polar spaces of constant curvature. On the other hand, is inherent need to dig deeper in trigonometry and geometry by their high applicability, because they are disciplines in which the properties of several special types of geometric figures are investigated [34]. For this purpose, it is important to generate spaces for reflection with respect to the theoretical basis, in order to be able to find relationships between the known theorems and other results that may arise from the continuous study of an area, particularly trigonometry.

4. Conclusions

The generation of theorems considered as a law of external nature [35], evidence of the commitment made by the teacher and the student during the academic process, this type of results shows that learning already there under the number of learned topics, but by the strategy of independent learning, which recognizes both the student and the teacher. Such changes in the teaching demands greater commitment by actors of the process, and as way to encourage meaningful learning. The importance of continuing to increase the theoretical tools of a certain area of knowledge, such as trigonometry, allows other researchers to continue their activities based on these results. We have shown that given an oblique triangle the value of any of its interior angles will always be less than any measure of the exterior angles other than its supplementary and in addition to this, another result shown as a corollary of the theorem stated in which we guarantee that if two interior angles are equal, then under the law of the sine, there will be a single triangle at most that satisfies the given basic conditions.

Reference

[1] Martinez F 2009 *Rev. Doc. Uni.* **10** 123-136
[2] Arbelaez R 2009 *Concepciones sobre competencias* (Bucaramanga: Cededuis)
[3] Acodesi 2002 *La formación integral y sus dimensiones* (Bogotá:Kimpres)

[4] Restrepo L 1996 *Proyecto de educación sexual* (Bogotá: Ministerio de Educación Nacional)
[5] Escobar M 2013 *Inv. And.* **15(27)** 733-735
[6] Mújica A 2012 *Rev. Doc. Uni.* **13** 201-216
[7] Maldonado M 2007 *Laur. Rev. Edu.* **13(23)** 263-278
[8] Perez M 2011 *Principios de aprendizaje* (Bucaramanga: Publicaciones Universidad Industrial de Santander)
[9] Martinez M and Esteban F 2005 *Rev. Esp. Ped.* **230** 63-73
[10] Grabovskij M and Kotel P 1971 *Edu. Stu. Math.* **3(2)** 147-160
[11] De Kee S, Mura R and Dionne J 1996 *For Lear. Math.* **16(2)** 19–27
[12] Maldonado E 2005 *Un análisis didáctico de la función trigonométrica* (México: Centro de Investigación y de Estudios Avanzados del Instituto Politécnico Nacional)
[13] Goel S and Elstak I 2015 *Geor. J. Sci.* **73** 136-146
[14] Farideh G and Sadighpour L 2012 *J. Oral. Impl.* **38** 289-290
[15] Upasani V, Chambers R, Dalal A, Shah S, Lehman R and Newton P 2009 *Spine* **34(17)** 1855–1862
[16] Lima T, Alves C, and Funayama C 2008 *J. Child. Neur.* **23** 1451-1454
[17] Van Wijck F, Pandyan A, Johnson G, and Barnes M 2001 *Neur. Neur. Rep.* **15(1)** 23-30
[18] Gosselin J, Gahagan S and Amiel-Tison C 2005 *Ment. Retard. Dev. Dis. Res. Rev.* **11(1)** 34-51
[19] Dargassies S. 1997 *Ment. Retard. Dev. Dis. Res. Rev.* (Buenos Aires: Editorial Médica Panamericana)
[20] Espinoza D, Florez R, Martinez A and Gnecchi J 2014 *J. Lap. & Adv. Sur. Tech.* **24** 1-8
[21] Ceylan A and Baycal O 2008 *Surv. Rev.* **40(308)** 195-205
[22] Ceylan A, Inal C, and Sanlioglu I 2005 *Modern height determination thechniques and comparison of accuracies* (Egipt: FIG Working Week)
[23] Gomes J 2006 *Determinação de desníveis de precisão utilizando estação total-dissertação de mestrado* (Brasil: Universidade Federal do Paraná)
[24] Perozzo Dos Santos D, Faggion P and Koenig L 2011 *Bol. Ciên. Geo.* **17(2)** 295-316
[25] Faggion P 2006 *Considerações sobre a instrumentação topográfica, curso de pós-graduação em ciências geodésicas* (Brasil:Universidade Federal do Paraná)
[26] Emmanouil E,Wei G, and Dai J 2016 *Robotica* **34** 2788–2805
[27] Skurnick R and Mohammad J 2006 *Math. Comp. Educ.* **40** 70-73
[28] Zill D and Dewar J 2012 *Algebra, trigonometria y geometria analitica* (Mexico: Mc Graw Hill)
[29] Ramirez A 2017 *Geometría analítica* (Mexico: Las prensas de ciencias)
[30] Páramo G and Benitez A 2015 *Ing. Indu.* **90** 43-48
[31] Kokkendorff L 2007 *J. Geom.* **86** 140-149
[32] Sher L and Sher D 2007 *Math. Comp. Educ.* **41** 168-171
[33] Karjanto N 2011 *Teac. Math. Appl.* **30(2)** 70-74
[34] Tarski A 1948 *A decision method for elementary algebra y geometry* (California: The Rand Corporation)
[35] Moritz R 1958 *On mathematics: A collection of witty* (New York: Dover publication)

Institute of Physics
Dirac House, Temple Back
Bristol BS1 6BE UK

ISSN: 1742-6588
ISBN 978-1-5108-6127-5

CURRAN ASSOCIATES INC.
proceedings.com

International Conference on Gravitation Cosmology and Continuum Mechanics 2016

Dedicated to the Centenary of Kirill Petrovich Staniukovich

Journal of Physics: Conference Series Volume 731

Moscow, Russia
3 - 4 March 2016

IOP
Institute of Physics